Michael V. Lurie

Modeling of Oil Product and Gas Pipeline Transportation

Related Titles

Olah, G. A., Goeppert, A., Prakash, G. K. S.

Beyond Oil and Gas: The Methanol Economy

2006
ISBN: 978-3-527-31275-7

Olah, G. A., Prakash, G. K. S. (eds.)

Carbocation Chemistry

2004
ISBN: 978-0-471-28490-1

Sheldrake, A. L.

Handbook of Electrical Engineering – For Practitioners in the Oil, Gas and Petrochemical Industry

2003
E-Book
ISBN: 978-0-470-86478-4

Michael V. Lurie

Modeling of Oil Product and Gas Pipeline Transportation

WILEY-BLACKWELL

WILEY-VCH Verlag GmbH & Co. KGaA

The Author

Prof. Dr. Michael V. Lurie
Russian State University
of Oil and Gas
Moscow, Russian Federation

Translation
Emmanuil G. Sinaiski
Leipzig, Germany

Cover Picture
Trans-Alaska Pipeline

Library of Congress Card No.: applied for
British Library Cataloguing-in-Publication Data
A catalogue record for this book is available from the British Library.

Bibliographic information published by the Deutsche Nationalbibliothek
Die Deutsche Nationalbibliothek lists this publication in the Deutsche Nationalbibliografie; detailed bibliographic data are available in the Internet at <http://dnb.d-nb.de>.

Printed in the Federal Republic of Germany
Printed on acid-free paper

Composition Laserwords, Chennai

Printing Strauss GmbH, Mörlenbach

Bookbinding Litges & Dopf GmbH, Heppenheim

ISBN: 978-3-527-40833-7

In memory of the Teacher – academician Leonid I. Sedov

Modeling of Oil Product and Gas Pipeline Transportation. Michael V. Lurie

ISBN: 978-3-527-40833-7

Foreword

This book is dedicated first and foremost to holders of a master's degree and postgraduate students of oil and gas institutes who have decided to specialize in the field of theoretical problems in the transportation of oil, oil products and gas. It contains methods of mathematical modeling of the processes taking place in pipelines when transporting these media.

By the term *mathematical model* is understood a system of mathematical equations in which framework a class of some processes could be studied. The solution of these equations provides values of parameters without carrying out model and, especially, full scale experiments.

Physical laws determining the dynamics of fluids and gases in pipes are presented. It is then shown how these laws are transformed into mathematical equations that are at the heart of one or another mathematical model. In the framework of each model, are formulated problems with the aim of investigating concrete situations. In doing so there are given methods of its solution.

The book is self-sufficient for studying the subject but the text is outlined in such a way that it impels the reader to address oneself to closer acquaintance of considered problem containing in special technical literature.

Professor Michael V. Lurie
Moscow

Modeling of Oil Product and Gas Pipeline Transportation. Michael V. Lurie

ISBN: 978-3-527-40833-7

Contents

Modeling of Oil Product and Gas Pipeline Transportation. Michael V. Lurie

ISBN: 978-3-527-40833-7

Preface

This book presents the fundamentals of the mathematical simulation of processes of pipeline transportation of oil, oil products and gas. It is shown how the basic laws of mechanics and thermodynamics governing the flow of fluids and gases in pipelines are transformed into mathematical equations which are the essence of a certain *mathematical model* and, in the framework of a given *physical problem*, appropriate *mathematical problems* are formulated to analyze concrete situations.

The book is suitable for graduate and postgraduate students of universities having departments concerned with oil and gas and to engineers and research workers specializing in pipeline transportation.

Beginners will find in this book a consecutive description of the theory and mathematical simulation methods of stationary and non-stationary processes occurring in pipelines. Engineers engaged in the design of and calculations on pipelines will find a detailed theoretical and practical text-book on the subject of their work. Graduate and postgraduate students and research workers will become acquainted with situations in the theory and methods in order to generalize and develop them in the future.

The author of the book, Professor Dr. M. Lurie, is a great authority in Russia in the field of the hydromechanics of oil and gas pipeline transportation.

Prof. Emmanuil Sinaiski
Leipzig

"... No human investigation could be referred to as true when it is not supported by mathematical proof"

Leonardo da Vinci

Modeling of Oil Product and Gas Pipeline Transportation. Michael V. Lurie

ISBN: 978-3-527-40833-7

List of Symbols

Symbol	Definition
a	radius of the flow core
a	dimensionless constant
a	parameter of the $(Q - \Delta H)$ characteristic
A	proportionality factor
A^+	value of parameter A to the left of the discontinuity front
A^-	value of parameter A to the right of the discontinuity front
$[A] = A^+ - A^-$	jump of parameter A at the discontinuity front
dA^{in}	elementary work of internal force
dA^{ex}	elementary work of external force
b	parameter of the $(Q - \Delta H)$ characteristic
c	velocity of wave propagation in a pipeline
c	sound velocity in gas
C_2	integration constant
C_f	friction factor
C_p	heat capacity at constant pressure
C_{Sh}	Chezy factor
C_v	heat capacity at constant volume
cP	centipoise, 0.01 P
cSt	centistokes, 0.01 St $= 10^{-6}$ m^2 s^{-1}
d	pipeline internal diameter
Δd	diameter increment
d_0	nominal internal diameter of pipeline; cylinder internal diameter
D	pipeline external diameter
D	velocity of hydraulic shock wave propagation in a pipeline
D	velocity of discontinuity front propagation in the positive direction of the x-axis
D_{im}	diameter of impeller
D_p	diameter of pump impeller
D_*	Joule–Thompson factor
e_{in}	internal energy density; specific internal energy
e_{kin}	kinetic energy density; specific kinetic energy

Modeling of Oil Product and Gas Pipeline Transportation. Michael V. Lurie

ISBN: 978-3-527-40833-7

E	elastic modulus in extension and compression, Young's modulus
E_{in}	internal energy
E_{kin}	kinetic energy
$Ei(z)$	Euler function
$\tilde{f}_1$	dimensionless factor
$f_{\tau}(Q)-$	friction force
F	restoring force
$\mathrm{d}F_{\mathrm{n}}$	elementary force
Fr	Froude number
g	acceleration due to gravity
g_0, g_1	dimensionless constants
h	piezometric head
$h(S)$	depth of pipeline cross-section filling with fluid
h_{c}	head losses in station communications
h_{cr}	critical depth
$h_{\pi.}$	head before PLP
h_{n}	normal depth of gravity flow in the pipe
H	head
ΔH	differential head
$\Delta H = F(Q)$	head-discharge $(Q - H)$ characteristic of a pump
H_1	hydraulic head
$\tilde{H}_1$	hydraulic head
He	Hedstroem number
i	hydraulic gradient
i_0	hydraulic gradient
I	momentum
I	Ilyushin number
I_1, I_2	Riemann invariants
J	gas enthalpy
k	factor of string elasticity
k	factor of power, Ostwald fluid
k	parameter of non-Newtonian fluid
κ	factor; heat-transfer factor; empirical factor $1/K$
κ	Karman constant
k	dimensionless constant
k	kinematic consistency
K	heat transfer factor
K	elastic modulus of fluid, Pa
K	factor of longitudinal mixing of oil product
l_{c}	length of the mixture region
L	length of a pipeline or a pipeline section
$\dot{M}$	mass flow rate
$\dot{M}_0$	initial mass flow rate
OPS	oil-pumping station

n	factor of power, Ostwald fluid
n	exponent in Ostwald rheological law
n	exponent
n	number of revolutions of centrifugal blower shaft
$\mathbf{n}$	unit normal vector
n_0	nominal number of revolutions of blower shaft
n_{in}	specific power of internal friction forces
N	power consumption, kW
N_{mech}	power of external mechanical devices
N_{us}	useful power of mechanical force acting on gas
N/ρ_e	specific power
p	pressure
Δp	difference between internal and external pressures, pressure drop
p_0	nominal pressure, initial pressure, normal pressure, pressure at the beginning of the pipeline section
p_{en}	pressure of gas at the entrance of compressor station and blower
p_{cr}	critical pressure
p_{ex}	external pressure; pressure at initial cross-section of the pipeline section
p_{in}	internal pressure
p_L	pressure at the end of the pipeline section
$p_{\text{l}}, p_{\text{e}}, p_{\pi}$	pressure at the pressure line of pumps (PLP)
p_{r}	reduced pressure
p_{st}	standard pressure, $p_{\text{st}} = 101\,325$ Pa
p_{u}	pressure before oil-pumping station
p_{u}	head before station
p_{v}	saturated vapor tension (pressure)
$[p]$	pressure jump
$[p_{\text{inc}}]$	incident pressure wave amplitude
$[p_{\text{refl}}]$	reflected pressure wave amplitude
$[p_{\text{trans}}]$	transmitted pressure wave amplitude
P	poise, 0.1 kg m s^{-1}
P_{s}	wetted perimeter
Pa	pascal (SI unit), kg m^{-1} s^{-2}
Pe	Peclet number
q_{h}	specific heat flux
q^{ex}	heat inflow ($q^{\text{ex}} > 0$) to gas; heat outflow ($q^{\text{ex}} < 0$) from gas
q_{M}	specific mass flow rate
q_{n}	external heat flux
Q	volume flow rate; fluid flow rate
Q_{e}	flow rate of gas at the entrance to the compressor station
Q_{k}	commercial flow rate of gas
Q_{M}	mass flow rate

Q_v	volume flow rate of gas at pipeline cross-section
r	radial coordinate
r_0	pipeline radius
R	gas constant ($R = R_0/\mu_g$)
R_0	universal gas constant
R_h	hydraulic radius
R_{im}	radius of the impeller
R_r	reduced gas constant
Re	Reynolds number
Re_{cr}	critical Reynolds number
Re_*	generalized Reynolds number
S	area of a cross-section; area of pipeline cross-section part filled with fluid
S_0	area of pipeline cross-section; nominal (basic) area
St	stokes, 10^{-4} m^2 s^{-1}
t	time
T	absolute temperature
T_0	nominal temperature; initial temperature; temperature of fluid at normal condition
T_{av}	average temperature over pipeline section length
T_{cr}	critical temperature
$T_{\mathcal{B}}$	temperature of gas at the entrance to the compressor station
T_{ex}	temperature of external medium
T_L	temperature at the end of pipeline section
T_m	mean temperature
T_r	reduced temperature
T_{st}	standard absolute temperature
$u(y)$	velocity distribution over cross-section
u_{max}	maximum value of velocity
u_w	fluid velocity at pipe wall
u_*	dynamic velocity
v	velocity averaged over cross-section
v	mean flow rate velocity
v_{cr}	critical velocity
V	volume
$[v]$	fluid velocity jump
w	acceleration
x	coordinate along the pipeline axis
x_1	coordinate of gravity flow section beginning
x_2	coordinate of gravity flow section end
y	coordinate transverse to the pipeline axis; direction of a normal to the elementary surface $d\sigma$
$z(x)$	elevation level of a pipeline cross-section x
$(z_1 - z_2)$	geometrical height differences of sections 1 and 2
Z	over-compressibility factor

Z_{av}	average over-compressibility factor
$Z = Z_r$	reduced gas over-compressibility
α	angle of inclination of the pipeline axis to the horizontal
α_κ, $\bar{\alpha}_\kappa$	factors
α_v	volume expansion factor
α_T	thermal expansion factor
β	compressibility factor
γ	adiabatic index
γ	ratio between the hydraulic gradient of pipeline section completely filled with fluid and the absolute value of the gravity flow section with slope α_p to the horizontal
$\dot{\gamma}$	shear rate, s^{-1}
δ	pipeline wall thickness
Δ	absolute equivalent roughness; roughness of wall surface
ε	relative roughness; compression ratio; thickness ratio
$\varsigma(t)$	local resistance factor
η	dimensionless radius
$\eta(\%)$	efficiency
θ	function of temperature; concentration; parameter of over-compressibility factor; parameter of state equation of real gas; concentration of anti-turbulent additive
λ	hydraulic resistance factor
λ_{eff}	effective factor of hydraulic resistance
μ	dynamic viscosity factor kg m^{-1} s^{-1}
μ_g	molar mass of gas; molecular weight
μ_t	turbulent dynamic viscosity
$\tilde{\mu}$	apparent viscosity of power Ostwald fluid
δ	kinematic viscosity factor m^2 s^{-1}
ν_0	kinematic viscosity factor at temperature T_0
ν_1	kinematic viscosity factor at temperature T_1
ν_P	Poisson ratio
ν_t	turbulent kinematic viscosity
ξ	factor of volumetric expansion, K^{-1}; self-similar coordinate; dimensionless coordinate
Π	dimensionless parameter; similarity criterion
$\Pi(x)$	initial pressure distribution
ρ	density
ρ^-, v^-, p^-, S^-	values of parameters before hydraulic shock wave
ρ^+, v^+, p^+, S^+	values of parameters after hydraulic shock wave
ρ_0	nominal density; fluid density at p_0; density of fluid under normal conditions
ρ_{st}	gas density under standard conditions
σ	area of suction branch pipe cross-section; hoop stress; degree of pipe filling; circumferential stress

$d\sigma$	elementary surface area; surface element
τ	tangential (shear) stress
	tangential friction stress
τ_0	critical (limit) shear stress
τ_w	tangential (shear) stress at the pipeline internal surface
υ	specific volume
υ_{cr}	critical specific volume
φ	angle of inclination of a straight line to the abscissa; central angle
$\Phi(x)$	initial fluid velocity distribution
ω	frequency of rotor rotation; angular velocity of impeller rotation

1
Fundamentals of Mathematical Modeling of One-Dimensional Flows of Fluid and Gas in Pipelines

1.1
Mathematical Models and Mathematical Modeling

Examination of phenomena is carried out with the help of *models*. Each model represents a definite schematization of the phenomenon taking into account not all the characteristic factors but some of them governing the phenomena and characterizing it from some area of interest to the researcher.

For example, to examine the motion of a body the *material point model* is often used. In such a model the dimensions of the body are assumed to be equal to zero and the whole mass to be concentrated at a point. In other words we ignore a lot of factors associated with body size and shape, the material from which the body is made and so on. The question is: to what extent would such a schematization be efficient in examining the phenomenon? As we all know such a body does not exist in nature. Nevertheless, when examining the motion of planets around the sun or satellites around the earth, and in many other cases, the material point model gives brilliant results in the calculation of the trajectories of a body under consideration.

In the examination of oscillations of a small load on an elastic spring we meet with greater schematization of the phenomenon. First the load is taken as a point mass m, that is we use the material point model, ignoring body size and shape and the physical and chemical properties of the body material. Secondly, the elastic string is also schematized by replacing it by the so-called *restoring force* $F = -k \cdot x$, where $x(t)$ is the deviation of the material point modeling the load under consideration from the equilibrium position and k is the factor characterizing the elasticity of the string. Here we do not take into account the physical-chemical properties of the string, its construction and material properties and so on. Further schematization could be done by taking into account the drag arising from the air flow around the moving load and the rubbing of the load during its motion along the guide.

The use of the differential equation

$$m\frac{d^2x}{dt^2} = -k \cdot x, \tag{1.1}$$

Modeling of Oil Product and Gas Pipeline Transportation. Michael V. Lurie

ISBN: 978-3-527-40833-7

expressing the second Newtonian law is also a schematization of the phenomenon, since the motion is described in the framework of Euclidian geometry which is the model of our space without taking into account the relativistic effects of the relativity theory.

The fact that the load motion can begin from an arbitrary position with an arbitrary initial velocity may be taken into account in the schematization by specifying *initial conditions* at

$$t = 0: \ x = x_0; v = \left(\frac{dx}{dt}\right)_0 = v_0. \tag{1.2}$$

Equation (1.1) represents the closed *mathematical model* of the considered phenomenon and when the initial conditions are included (1.2) this is the *concrete mathematical model* in the framework of this model. In the given case we have the so-called *initial value (Cauchy) problem* allowing an exact solution. This solution permits us to predict the load motion at instants of time $t > 0$ and by so doing to discover regularities of its motion that were not previously evident. The latest circumstance contains the whole meaning and purpose of mathematical models.

It is also possible of course to produce another more general schematization of the same phenomenon which takes into account a great number of characteristic factors inherent to this phenomenon, that is, it is possible, in principle, to have another more general model of the considered phenomenon.

This raises the question, how can one tell about the correctness or incorrectness of the phenomenon schematization when, from the logical point of view, both schematizations (models) are consistent? The answer is: only from results obtained in the framework of these models. For example, the above-outlined model of load oscillation around an equilibrium position allows one to calculate the motion of the load as

$$x(t) = x_0 \cdot \cos\left(\sqrt{\frac{k}{m}} \cdot t\right) + \sqrt{\frac{m}{k}} \cdot v_0 \cdot \sin\left(\sqrt{\frac{k}{m}} \cdot t\right)$$

having undamped periodic oscillations. How can one evaluate the obtained result? On the one hand there exists a time interval in the course of which the derived result accords well with the experimental data. Hence the model is undoubtedly *correct and efficient*. On the other hand the same experiment shows that oscillations of the load are gradually damping in time and come to a stop. This means that the model (1.1) and the problem (1.2) do not take into account some factors which could be of interest for us, and the accepted schematization is inadequate.

Including in the number of forces acting on the load additional forces, namely the forces of dry $-f_0 \cdot \text{sign}(\dot{x})$ and viscous $-f_1 \cdot \dot{x}$ friction (where the symbol $\text{sign}(\dot{x})$ denotes the function $\dot{x}$– sign equal to 1, at $\dot{x} > 0$; equal to -1, at $\dot{x} < 0$ and equal to 0, at $\dot{x} = 0$), that is using the equation

$$m\frac{d^2x}{dt^2} = -k \cdot x - f_0 \cdot \text{sign}(\dot{x}) - f_1 \cdot \dot{x} \tag{1.3}$$

instead of Eq. (1.1), one makes the schematization (model) more complete. Therefore it adequately describes the phenomenon.

But even the new model describes only *approximately* the model under consideration. In the case when the size and shape of the load strongly affect its motion, the motion itself is not one-dimensional, the forces acting on the body have a more complex nature and so on. Thus it is necessary to use more complex schematizations or in another words to exploit more complex models. Correct schematization frequently represents a challenging task, requiring from the researcher great experience, intuition and deep insight into the phenomenon to be studied (Sedov, 1965).

Of special note is the *continuum model,* which occupies a highly important place in the following chapters. It is known that all media, including liquids and gases, comprise a great collection of different atoms and molecules in permanent heat motion and with complex interactions. By molecular interactions we mean such properties of real media as compressibility, viscosity, heat conductivity, elasticity and others. The complexity of these processes is very high and the governing forces are not always known. Therefore such seemingly natural investigation of medium motion through a study of discrete molecules is absolutely unacceptable.

One of the general schematization methods for fluid, gas and other deformable media motion is based on the continuum model. Because each macroscopic volume of the medium under consideration contains a great number of molecules the medium could be approximately considered as if it fills the space continuously. Oil, oil products, gas, water or metals may be considered as a medium continuously filling one or another region of the space. That is why *a system of material points continuously filling a part of space is called a continuum.*

Replacement of a real medium consisting of separate molecules by a continuum represents of course a schematization. But such a schematization has proved to be very convenient in the use of the mathematical apparatus of continuous functions and, as was shown in practice, it is quite sufficient for studying the overwhelming majority of observed phenomena.

1.1.1 Governing Factors

In the examination of different phenomena the researcher is always restricted by a finite number of parameters called *governing factors (parameters)* within the limits of which the investigation is being studied. This brings up the question: How to reveal the system of governing parameters?

It could be done for example by formulating the problem mathematically or, in other words, by building a mathematical model of the considered phenomenon as was demonstrated in the above-mentioned example. In this problem the governing parameters are:

$$x,\ t,\ m,\ k,\ f_0,\ f_1,\ x_0,\ v_0.$$

But, in order to determine the system of governing parameters, there is no need for mathematical schematization of the process. It is enough to be guided, as has already been noted, by experience, intuition and understanding of the mechanism of the phenomenon.

Let us investigate the decrease in a parachutist's speed v in the air when his motion can be taken as steady. Being governed only by intuition it is an easy matter to assume the speed to be dependent on the mass of the parachutist m, acceleration due to g, the diameter of the parachute canopy D, the length L of its shroud and the air density ρ. The viscosity of the air flowing around the parachute during its descent can be taken into account or ignored since the force of viscous friction is small compared to parachute drag. Both cases represent only different schematizations of the phenomenon.

So the function sought could be assumed to have the following general form $v = f(m, g, D, L, \rho)$. Then the governing parameters are:

$$m,\ g,\ D,\ L,\ \rho.$$

The use of dimensional theory permits us to rewrite the formulated dependence in invariant form, that is, independent of the system of measurement units (see Chapters 6 and 7)

$$\frac{v}{\sqrt{gD}} = \tilde{f}\left(\frac{m}{\rho D^3}, \frac{L}{D}\right), \Rightarrow v = \sqrt{gD}\cdot\tilde{f}\left(\frac{m}{\rho D^3}, \frac{L}{D}\right).$$

Thus, among five governing parameters there are only two independent dimensionless combinations, $m/\rho D^3$ and L/D, defining the sought-for dependence.

1.1.2
Schematization of One-Dimensional Flows of Fluids and Gases in Pipelines

In problems of oil and gas transportation most often schematization of the flow process under the following conditions is used:

- oil, oil product and gas are considered as a continuum continuously filling the whole cross-section of the pipeline or its part;
- the flow is taken as one-dimensional, that is all governing parameters depend only on one space coordinate x measured along the pipeline axis and, in the general case, on time t;
- the governing parameters of the flow represent values of the corresponding physical parameters averaged over the pipeline cross-section;

- the profile of the pipeline is given by the dependence of the height of the pipeline axis above sea level on the linear coordinate $z(x)$;
- the area S of the pipeline cross-section depends, in the general case, on x and t. If the pipeline is assumed to be undeformable, then $S = S(x)$. If the pipeline has a constant diameter, then $S(x) = S_0 = \text{const.}$;
- the most important parameters are:
  $\rho(x, t)$ – density of medium to be transported, kg m^{-3};
  $v(x, t)$ – velocity of the medium, m s^{-1};
  $p(x, t)$ – pressure at the pipeline axis, Pa = N m^{-2};
  $T(x, t)$ – temperature of the medium to be transported, degrees;
  $\tau(x, t)$ – shear stress (friction force per unit area of the pipeline internal surface), Pa = N m^{-2};
  $Q(x, t) = vS$ – volume flow rate of the medium, m^3 s^{-1};
  $\dot{M}(x, t) = \rho v S$ – mass flow rate of the medium, kg s^{-1} and other.

Mathematical models of fluid and gas flows in the pipeline are based on the fundamental laws of physics (mechanics and thermodynamics) of a continuum, modeling a real fluid and a real gas.

1.2 Integral Characteristics of Fluid Volume

In what follows one needs the notion of *movable fluid volume* of the continuum in the pipeline. Let, at some instant of time, an arbitrary volume of the medium be transported between cross-sections x_1 and x_2 of the pipeline (Figure 1.1).

If the continuum located between these two cross-sections is identified with a system of material points and track is kept of its displacement in time, the boundaries x_1 and x_2 become dependent on time and, together with the pipeline surface, contain one and the same material points of the continuum. This volume of the transported medium is called the *movable fluid volume* or *individual volume*. Its special feature is that it always *consists of the same particles of the continuum* under consideration. If, for example, the transported medium is incompressible and the pipeline is non-deformable, then $S = S_0 = \text{const.}$ and the difference between the demarcation boundaries $(x_2 - x_1)$ defining the length of the fluid volume remains constant.

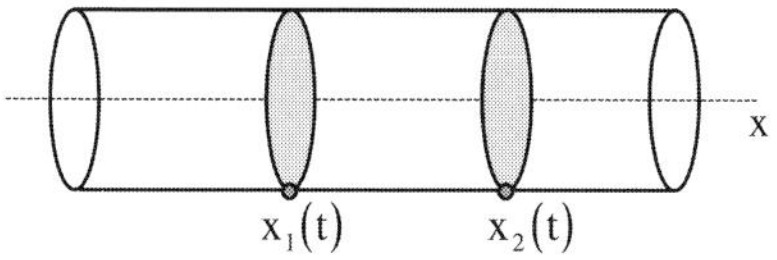

Figure 1.1 Movable fluid volume of the continuum.

Exploiting the notion of fluid or individual volume of the transported medium in the pipeline one can introduce the following integral quantities:

$$M = \int_{x_1(t)}^{x_2(t)} \rho(x,t) \cdot S(x,t)\,dx \text{ -- mass of fluid volume (kg);}$$

$$I = \int_{x_1(t)}^{x_2(t)} \rho(x,t) \cdot v(x,t) \cdot S(x,t)\,dx \text{ -- momentum of fluid volume}$$

$$(\text{kg m s}^{-1});$$

$$E_{\text{kin}} = \int_{x_1(t)}^{x_2(t)} \alpha_k \frac{\rho v^2}{2} S(x,t)\,dx \text{ -- kinetic energy of the fluid volume } (J),$$

where α_k is the factor;

$$E_{\text{in}} = \int_{x_1(t)}^{x_2(t)} \rho(x,t) \cdot e_{\text{in}}(x,t) \cdot S(x,t)\,dx \text{ -- internal energy of the fluid}$$

volume, where e_{in} is the density of the internal energy (J kg^{-1}), that is the internal energy per unit mass.

These quantities model the mass, momentum and energy of a material point system.

Since the main laws of physics are often formulated as connections between physical quantities and the rate of their change in time, we ought to adduce the rule of integral quantity differentiation with respect to time. The symbol of differentiation $\mathrm{d}()/\mathrm{d}t$ denotes the total derivative with respect to time, associated with *individual* particles of a continuum whereas the symbol $\partial()/\partial t$ denotes the local derivative with respect to time, that is the derivative of a flow parameter with respect to time at a *given space point*, e.g. $x = \text{const}$. The local derivative with respect to time gives the rate of flow parameter change at a given cross-section of the flow while, at two consecutive instances of time, different particles of the continuum are located in this cross-section.

The total derivative with respect to time is equal to

$$\frac{\mathrm{d}}{\mathrm{d}t} \int_{x_1(t)}^{x_2(t)} A(x,t) \cdot S(x,t)\,\mathrm{d}x.$$

From mathematical analysis it is known how an integral containing a parameter, in the considered case it is t, is differentiated with respect to this parameter, when the integrand and limits of integration depend on this parameter. We have

$$\frac{\mathrm{d}}{\mathrm{d}t} \int_{x_1(t)}^{x_2(t)} A(x,t) \cdot S(x,t)\,\mathrm{d}x = \int_{x_1(t)}^{x_2(t)} \frac{\partial}{\partial t}[A(x,t) \cdot S(x,t)]\,\mathrm{d}x$$
$$+ A(x,t) \cdot S(x,t)|_{x_2(t)} \cdot \frac{\mathrm{d}x_2}{\mathrm{d}t} - A(x,t) \cdot S(x,t)|_{x_1(t)} \cdot \frac{\mathrm{d}x_1}{\mathrm{d}t}.$$

First, at frozen upper and lower integration limits, the integrand is differentiated (the derivative being local) and then the integrand calculated at the upper and lower integration limits is multiplied by the rates of change of these limits dx_2/dt and dx_1/dt, the first term having been taken with a plus sign and the second with a minus sign (see Appendix B).

For the case of the fluid volume of the medium the quantities dx_2/dt and dx_1/dt are the corresponding velocities $v_2(t)$ and $v_1(t)$ of the medium in the left and right cross-sections bounding the considered volume. Hence

$$\frac{d}{dt}\int_{x_1(t)}^{x_2(t)} A(x,t)\cdot S(x,t)\,dx = \int_{x_1(t)}^{x_2(t)} \frac{\partial}{\partial t}[A(x,t)\cdot S(x,t)]\,dx$$
$$+ A(x,t)\cdot v(x,t)\cdot S(x,t)|_{x_2(t)} - A(x,t)\cdot v(x,t)\cdot S(x,t)|_{x_1(t)}.$$

If, in addition, we take into account the well-known Newton–Leibniz formula, according to which

$$A(x,t)\cdot v(x,t)\cdot S(x,t)|_{x_2(t)} - A(x,t)\cdot v(x,t)\cdot S(x,t)|_{x_1(t)}$$
$$= \int_{x_1(t)}^{x_2(t)} \frac{\partial}{\partial x}[A(x,t)\cdot v(x,t)\cdot S(x,t)]\,dx,$$

we obtain

$$\frac{d}{dt}\int_{x_1(t)}^{x_2(t)} A(x,t)\cdot S(x,t)\,dx = \int_{x_1(t)}^{x_2(t)} \left(\frac{\partial AS}{\partial t} + \frac{\partial ASv}{\partial x}\right) dx. \tag{1.4}$$

1.3
The Law of Conservation of Transported Medium Mass. The Continuity Equation

The density $\rho(x,t)$, the velocity of the transported medium $v(x,t)$ and the area of the pipeline cross-section $S(x,t)$ cannot be chosen arbitrarily since their values define the enhancement or reduction of the medium mass in one or another place of the pipeline. Therefore the first equation would be obtained when the transported medium is governed by the mass conservation law

$$\frac{d}{dt}\int_{x_1(t)}^{x_2(t)} \rho(x,t)\cdot S(x,t)\,dx = 0, \tag{1.5}$$

This equation should be obeyed for any fluid particle of the transported medium, that is for any values $x_1(t)$ and $x_2(t)$.

Applying to Eq. (1.4) the rule (1.5) of differentiation of integral quantity with regard to fluid volume, we obtain

$$\int_{x_1(t)}^{x_2(t)} \left(\frac{\partial \rho S}{\partial t} + \frac{\partial \rho v S}{\partial x}\right) dx = 0.$$

Since the last relation holds for arbitrary integration limits we get the following differential equation

$$\frac{\partial \rho S}{\partial t} + \frac{\partial \rho v S}{\partial x} = 0, \tag{1.6}$$

which is called *continuity equation* of the transported medium in the pipeline.

If the flow is stationary, that is the local derivative with respect to time is zero ($\partial()/\partial t = 0$), the last equation is simplified to

$$\frac{\mathrm{d}\rho v S}{\mathrm{d}x} = 0 \Rightarrow \dot{M} = \rho v S = \text{const.} \tag{1.7}$$

This means that in stationary flow the mass flow rate $\dot{M}$ is constant along the pipeline.

If we ignore the pipeline deformation and take $S(x) \cong S_0 = \text{const.}$, from Eq. (1.7) it follows that $\rho v = \text{const.}$ From this follow two important consequences:

1. In the case of a homogeneous incompressible fluid (sometimes oil and oil product can be considered as such fluids) $\rho \cong \rho_0 = \text{const.}$ and the flow velocity $v(x) = \text{const.}$ Hence the *flow velocity of a homogeneous incompressible fluid in a pipeline of constant cross-section does not change along the length of the pipeline.*

Example. The volume flow rate of the oil transported by a pipeline with diameter $D = 820$ mm and wall thickness $\delta = 8$ mm is 2500 $\mathrm{m^3\,h^{-1}}$. It is required to find the velocity v of the flow.

Solution. The internal diameter d of the oil pipeline is equal to

$$d = D - 2\delta = 0.82 - 2 \cdot 0.008 = 0.804\ \mathrm{m};$$

$$v = 4Q/\pi d^2 = \text{const.}$$

$$v = 4 \cdot 2500/(3600 \cdot 3.14 \cdot 0.804^2) \cong 1.37\ \mathrm{m\,s^{-1}}.$$

2. In the case of a compressible medium, e.g. a gas, the density $\rho(x)$ changes along the length of pipeline section under consideration. Since the density is as a rule connected with pressure, this change represents a monotonic function decreasing from the beginning of the section to its end. Then from the condition $\rho v = \text{const.}$ it follows that the velocity $v(x)$ of the flow also increases monotonically from the beginning of the section to its end. Hence the velocity of the *gas flow in a pipeline with constant diameter increases from the beginning of the section between compressor stations to its end.*

Example. The mass flow rate of gas transported along the pipeline ($D = 1020$ mm, $\delta = 10$ mm) is 180 $\mathrm{kg\,s^{-1}}$. Find the velocity of the gas flow

v_1 at the beginning and v_2 at the end of the gas-pipeline section, if the density of the gas at the beginning of the section is 45 kg m^{-3} and at the end is 25 kg m^{-3}.

Solution. $v_1 = \dot{M}/(\rho_1 S) = 4 \cdot 180/(45 \cdot 3.14 \cdot 1^2) \cong 5.1$ m s^{-1}; $v_2 = \dot{M}/(\rho_2 S) = 4 \cdot 180/(25 \cdot 3.14 \cdot 1^2) \cong 9.2$ m s^{-1}, that is the gas flow velocity is enhanced by a factor 1.8 towards the end as compared with the velocity at the beginning.

1.4 The Law of Change in Momentum. The Equation of Fluid Motion

The continuity equation (1.6) contains several unknown functions, hence the use of only this equation is insufficient to find each of them. To get additional equations we can use, among others, the equation of the change in momentum of the system of material points comprising the transported medium. This law expresses properly the second Newton law applied to an arbitrary fluid volume of transported medium

$$\frac{\mathrm{d}I}{\mathrm{d}t} = \frac{\mathrm{d}}{\mathrm{d}t}\int_{x_1(t)}^{x_2(t)} v \cdot \rho S \,\mathrm{d}x = (p_1 S_1 - p_2 S_2) + \int_{x_1(t)}^{x_2(t)} p \frac{\partial S}{\partial x}\,\mathrm{d}x - \int_{x_1(t)}^{x_2(t)} \pi d \cdot \tau_w \,\mathrm{d}x - \int_{x_1(t)}^{x_2(t)} \rho g \sin\alpha(x) \cdot S \,\mathrm{d}x. \tag{1.8}$$

On the left is the total derivative of the fluid volume momentum of the transported medium with respect to time and on the right the sum of all external forces acting on the considered volume.

The first term on the right-hand side of the equation gives the difference in pressure forces acting at the ends of the single continuum volume. The second term represents the axial projection of the reaction force from the lateral surface of the pipe (this force differs from zero when $S \neq$ const.). The third term defines the friction force at the lateral surface of the pipe (τ_w is the shear stress at the pipe walls, that is the friction force per unit area of the pipeline internal surface, Pa). The fourth term gives the *sliding component* of the gravity force ($\alpha(x)$ is the slope of the pipeline axis to the horizontal, $\alpha > 0$ for ascending sections of the pipeline; $\alpha < 0$ for descending sections of the pipeline; g is the acceleration due to gravity).

Representing the pressure difference in the form of an integral over the length of the considered volume

$$p_1 S_1 - p_2 S_2 = -\int_{x_1(t)}^{x_2(t)} \frac{\partial pS}{\partial x}\,\mathrm{d}x$$

and noting that

$$-\int_{x_1(t)}^{x_2(t)} \frac{\partial pS}{\partial x}\,dx + \int_{x_1(t)}^{x_2(t)} p\frac{\partial S}{\partial x}\,dx = -\int_{x_1(t)}^{x_2(t)} S\frac{\partial p}{\partial x}\,dx,$$

we obtain the following equation

$$\frac{d}{dt}\int_{x_1(t)}^{x_2(t)} \rho v S\,dx = \int_{x_1(t)}^{x_2(t)} \left(-S\frac{\partial p}{\partial x} - S\cdot\frac{4}{d}\tau_w - S\rho g\sin\alpha(x)\right) dx.$$

Now applying to the left-hand side of this equation the differentiation rule of fluid volume

$$\int_{x_1(t)}^{x_2(t)} \left(\frac{\partial \rho v S}{\partial t} + \frac{\partial \rho v^2 S}{\partial x}\right) dx$$
$$= \int_{x_1(t)}^{x_2(t)} \left(-S\frac{\partial p}{\partial x} - S\cdot\frac{4}{d}\tau_w - S\rho g\sin\alpha(x)\right) dx.$$

As far as the limits of integration in the last relation are arbitrary one can discard the integral sign and get the differential equation

$$\frac{\partial \rho v S}{\partial t} + \frac{\partial \rho v^2 S}{\partial x} = S\cdot\left(-\frac{\partial p}{\partial x} - \frac{4}{d}\tau_w - \rho g\sin\alpha(x)\right). \tag{1.9}$$

If we represent the left-hand side of this equation in the form

$$v\left(\frac{\partial \rho S}{\partial t} + \frac{\partial \rho v S}{\partial x}\right) + \rho S\left(\frac{\partial v}{\partial t} + v\frac{\partial v}{\partial x}\right)$$

and take into account that in accordance with the continuity equation (1.6) the expression in the first brackets is equal to zero, the resulting equation may be written in a more simple form

$$\rho\left(\frac{\partial v}{\partial t} + v\frac{\partial v}{\partial x}\right) = -\frac{\partial p}{\partial x} - \frac{4}{d}\tau_w - \rho g\sin\alpha(x). \tag{1.10}$$

The expression in brackets on the left-hand side of Eq. (1.10) represents the total derivative with respect to time, that is the particle acceleration

$$w = \frac{dv}{dt} = \frac{\partial v}{\partial t} + v\frac{\partial v}{\partial x}. \tag{1.11}$$

Now the meaning of Eq. (1.10) becomes clearer: the product of unit volume mass of transported medium and its acceleration is equal to the sum of all forces acting on the medium, namely pressure, friction and gravity forces. So Eq. (1.10) expresses the Newton's Second Law and can therefore also be called the *flow motion equation*.

Remark. *about the connection between total and partial derivatives with respect to time.* The acceleration $w = dv/dt$ is a total derivative with respect to time (symbol $d()/dt$), since we are dealing with the velocity differentiation of one and the same fixed particle of the transported medium moving from one cross-section of the pipeline to another one, whereas the partial derivative with respect to time (symbol $\partial()/\partial t$) has the meaning of velocity differentiation at a given place in space, that is at a constant value of x. Thus such a derivative gives the change in velocity of different particles of the transported medium entering a given cross-section of the pipeline.

Let a particle of the medium at the instant of time t be in the cross-section x of the pipeline and so have velocity $v(x, t)$. In the next instant of time $t + \Delta t$ this particle will transfer to the cross-section $x + \Delta x$ and will have velocity $v(x + \Delta x, t + \Delta t)$. The acceleration w of this particle is defined as the limit

$$w = \frac{dv}{dt} = \lim_{\Delta t \Rightarrow 0} \frac{v(x + \Delta x, t + \Delta t) - v(x, t)}{\Delta t} = \left.\frac{\partial v}{\partial t}\right|_x + \left.\frac{\partial v}{\partial x}\right|_t \cdot \frac{dx}{dt}.$$

Since $dx/dt = v(x, t)$ is the velocity of the considered particle, from the last equality it follows that

$$\frac{dv}{dt} = \frac{\partial v}{\partial t} + v \cdot \frac{\partial v}{\partial x}. \tag{1.12}$$

A similar relation between the total derivative (d/dt), or as it is also called the individual or Lagrangian derivative, and the partial derivative $(\partial/\partial t)$, or as it is also called the local or Eulerian derivative, has the form (1.12) no matter whether the case in point is velocity or any other parameter $A(x, t)$

$$\frac{dA(x, t)}{dt} = \frac{\partial A(x, t)}{\partial t} + v \cdot \frac{\partial A(x, t)}{\partial x}.$$

1.5 The Equation of Mechanical Energy Balance

Consider now what leads to the use of the mechanical energy change law as applied to the system of material points representing a fluid particle of the transported medium. This law is written as:

$$\frac{dE_{kin}}{dt} = \frac{dA^{ex}}{dt} + \frac{dA^{in}}{dt} \tag{1.13}$$

that is the change in kinetic energy of a system of material points dE_{kin} is equal to the sum of the work of the external dA^{ex} and internal dA^{in} forces acting on the points of this system.

We can calculate separately the terms of this equation but first we should define more exactly what meant by the kinetic energy E_{kin}. If the transported medium moves in the pipeline as a piston with equal velocity $v(x, t)$ over the cross-section then the kinetic energy would be expressed as the integral

$$E_{kin} = \int_{x_1(t)}^{x_2(t)} \frac{\rho v^2}{2} S\, dx.$$

But, in practice, such a schematization is too rough because, as experiments show, the velocity of the separate layers of the transported medium (fluid or gas) varies over the pipe cross-section. At the center of the pipe it reaches the greatest value, whereas as the internal surface of the pipe is approached the velocity decreases and at the wall itself it is equal to zero. Furthermore, if at a small velocity of the fluid the flow regime is *laminar*, with an increase in velocity the laminar flow changes into a *turbulent* one (pulsating and mixing flow) and the velocities of the separate particles differ significantly from the average velocity v of the flow. That is why models of the flow are, as a rule, constructed with regard to the difference in flow velocity from the average velocity over the cross-section.

The true velocity u of a particle of the transported medium is given as the sum $u = v + \Delta u$ of the average velocity over the cross-section $v(x, t)$ and the additive one (deviation) Δu representing the difference between the true velocity and the average one. The average value of this additive $\overline{\Delta u}$ is equal to zero, but the root-mean-square (rms) value of the additive $\overline{(\Delta u)^2}$ is non-vanishing. The deviation characterizes the kinetic energy of the relative motion of the continuum particle in the pipeline cross-section. Then the kinetic energy of the transported medium unit mass e_{kin} may be presented as the sum of two terms

$$e_{\text{kin}} = \frac{v^2}{2} + \frac{\overline{(\Delta u)^2}}{2}$$

namely the kinetic energy of the center of mass of the considered point system and the kinetic energy of the motion of these points relative to the center of mass. If the average velocity $v \neq 0$, then

$$\frac{\rho v^2}{2} + \frac{\rho\overline{(\Delta u)^2}}{2} = \frac{\rho v^2}{2} \cdot \left(1 + \frac{\overline{(\Delta u)^2}}{v^2}\right) = \alpha_{\text{k}} \cdot \frac{\rho v^2}{2}$$

where $\alpha_k = 1 + \overline{(\Delta u)^2}/v^2 > 1$. For laminar flow $\alpha_{\text{k}} = 4/3$, while for turbulent flow the value of α_{k} lies in the range 1.02–1.05.

Remark. *It should be noted that in one-dimensional theory, as a rule, the cases* $v = 0$ *and* $\overline{(\Delta u)^2} \neq 0$ *are not considered.*

With regard to the introduced factor the kinetic energy of any movable volume of transported medium may be represented as

$$E_{\text{kin}} = \int_{x_1(t)}^{x_2(t)} \alpha_{\text{k}} \cdot \frac{\rho v^2}{2} \cdot S\,\text{d}x.$$

Let us turn now to the calculation of the terms in the mechanical energy equation (1.13). Let us calculate first the change in kinetic energy

$$\frac{\text{d}E_{\text{kin}}}{\text{d}t} = \frac{\text{d}}{\text{d}t}\left(\int_{x_1(t)}^{x_2(t)} \alpha_{\text{k}} \cdot \frac{\rho v^2}{2} S\,\text{d}x\right).$$

Employing the rule of integral quantity integration with reference to the fluid volume, that is an integral with variable integration limits, we get

$$\frac{\mathrm{d}E_{\mathrm{kin}}}{\mathrm{d}t} = \int_{x_1(t)}^{x_2(t)} \left[\frac{\partial}{\partial t}\left(\alpha_{\mathrm{k}} \cdot \frac{\rho v^2}{2} S\right) + \frac{\partial}{\partial x}\left(\alpha_{\mathrm{k}} \cdot \frac{\rho v^2}{2} S \cdot v\right)\right] \mathrm{d}x.$$

The work of the external forces (in this case they are the forces of pressure and gravity), including also the work of external mechanical devices, e.g. pumps if such are used, is equal to

$$\begin{aligned}\frac{\mathrm{d}A^{\mathrm{ex}}}{\mathrm{d}t} &= (p_1 S v_1 - p_2 S v_2) - \int_{x_1(t)}^{x_2(t)} \rho g \sin\alpha \cdot v \cdot S\,\mathrm{d}x + N_{\mathrm{mech}} \\ &= -\int_{x_1(t)}^{x_2(t)} \frac{\partial}{\partial x}(pSv)\,\mathrm{d}x - \int_{x_1(t)}^{x_2(t)} \rho g \sin\alpha \cdot v \cdot S\,\mathrm{d}x + N_{\mathrm{mech}}.\end{aligned}$$

The first term on the right-hand side of the last expression gives the work performed in unit time or, more precisely, the power of the pressure force applied to the initial and end cross-sections of the detached volume. The second term gives the power of the gravity force and the third term N_{mech} the power of the external mechanical devices acting on the transported medium volume under consideration.

The work of the internal forces (pressure and internal friction) executed in unit time is given by

$$\frac{\mathrm{d}A^{\mathrm{in}}}{\mathrm{d}t} = \int_{x_1(t)}^{x_2(t)} p\frac{\partial(Sv)}{\partial x}\,\mathrm{d}x + \int_{x_1(t)}^{x_2(t)} n^{in} \cdot \rho S\,\mathrm{d}x.$$

The first term on the right-hand side gives the work of the pressure force in unit time, that is the power, for compression of the particles of the medium, the factor $\partial(Sv)/\partial x \cdot \mathrm{d}x$ giving the rate of elementary volume change. The second term represents the power of the internal friction forces, that is the forces of mutual friction between the internal layers of the medium, n^{in} denoting specific power, that is per unit mass of the transported medium. In what follows it will be shown that this quantity characterizes the amount of mechanical energy converting into heat per unit time caused by mutual internal friction of the transported particles of the medium.

Gathering together all the terms of the mechanical energy equation we get

$$\begin{aligned}&\int_{x_1(t)}^{x_2(t)} \left[\frac{\partial}{\partial t}\left(\alpha_{\mathrm{k}} \cdot \frac{\rho v^2}{2} S\right) + \frac{\partial}{\partial x}\left(\alpha_{\mathrm{k}} \cdot \frac{v^2}{2} \rho v S\right)\right] \mathrm{d}x \\ &= -\int_{x_1(t)}^{x_2(t)} \rho S v \left[\left(\frac{1}{\rho}\frac{\partial p}{\partial x}\right) + g\sin\alpha\right] \mathrm{d}x + \int_{x_1(t)}^{x_2(t)} n^{\mathrm{in}} \cdot \rho S\,\mathrm{d}x + N_{\mathrm{mech}}.\end{aligned}$$

If the transported medium is barotropic, that is the pressure in it depends only on the density $p = p(\rho)$, one can introduce a function $P(\rho)$ of the pressure such that $dP = dp/\rho$, $P(\rho) = \int dp/\rho$ and $\frac{1}{\rho}\frac{\partial p}{\partial x} = \frac{\partial P(\rho)}{\partial x}$. If, moreover, we take into account the equality $\sin\alpha(x) = \partial z/\partial x$, where the function $z(x)$ is referred to as the pipeline profile, the last equation could be rewritten in the simple form

$$\int_{x_1(t)}^{x_2(t)} \left[\rho S \frac{\partial}{\partial t}\left(\frac{\alpha_k v^2}{2}\right) + \rho v S \frac{\partial}{\partial x}\left(\frac{\alpha_k v^2}{2} + P(\rho) + gz\right)\right] dx$$
$$= \int_{x_1(t)}^{x_2(t)} n^{\text{in}} \cdot \rho S \, dx + N_{\text{mech}}. \tag{1.14}$$

If we assume that in the region $[x_1(t), x_2(t)]$ external sources of mechanical energy are absent. Then $N_{\text{mech}} = 0$ and we can go from the integral equality (1.14) to a differential equation using, as before, the condition of arbitrariness of integration limits $x_1(t)$ and $x_2(t)$ in Eq. (1.14). Then the sign of the integral can be omitted and the corresponding differential equation is

$$\rho S \frac{\partial}{\partial t}\left(\frac{\alpha_k v^2}{2}\right) + \rho v S \frac{\partial}{\partial x}\left(\frac{\alpha_k v^2}{2} + P(\rho) + gz\right) = \rho S \cdot n^{\text{in}} \tag{1.15}$$

or

$$\frac{\partial}{\partial t}\left(\frac{\alpha_k v^2}{2}\right) + v \cdot \frac{\partial}{\partial x}\left(\frac{\alpha_k v^2}{2} + \int \frac{dp}{\rho} + gz\right) = n^{\text{in}}. \tag{1.16}$$

This is the sought differential equation expressing the law of mechanical energy change. It should be emphasized that this equation is not a consequence of the motion equation (1.10). It represents an independent equation for modeling one-dimensional flows of a transported medium in the pipeline.

If we divide both parts of Eq. (1.16) by g we get

$$\frac{\partial}{\partial t}\left(\frac{\alpha_k v^2}{2g}\right) + v \cdot \frac{\partial}{\partial x}\left(\frac{\alpha_k v^2}{2g} + \int \frac{dp}{\rho g} + z\right) = \frac{n^{\text{in}}}{g}.$$

The expression

$$H = \frac{\alpha_k v^2}{2g} + \int \frac{dp}{\rho g} + z \tag{1.17}$$

in the derivative on the left-hand side of the last equation has the dimension of length and is called the *total head*. The total head at the pipeline cross-section x consists of the *kinetic head (dynamic pressure)* $\alpha_k v^2/2g$, the *piezometric head* $\int dp/\rho g$ and the *geometric head* z. The concept of head is very important in the calculation of processes occurring in pipelines.

1.5.1 Bernoulli Equation

In the case of stationary flow of a *barotropic fluid or gas* in the pipeline the derivative $\partial()/\partial t = 0$, hence the following ordinary differential equations apply

$$v\frac{\mathrm{d}}{\mathrm{d}x}\left(\frac{\alpha_{\mathrm{k}}v^2}{2g} + \int\frac{\mathrm{d}p}{\rho g} + z\right) = \frac{n^{\mathrm{in}}}{g}$$

or

$$\frac{\mathrm{d}}{\mathrm{d}x}\left(\frac{\alpha_{\mathrm{k}}v^2}{2g} + \int\frac{\mathrm{d}p}{\rho g} + z\right) = \frac{n^{\mathrm{in}}}{gv} = i, \tag{1.18}$$

where i denotes the dimensional quantity n^{in}/gv called the *hydraulic gradient*

$$i = \frac{\mathrm{d}H}{\mathrm{d}x} = \frac{n^{\mathrm{in}}}{gv}.$$

Thus the hydraulic gradient, defined as the pressure loss per unit length of the pipeline, is proportional to the dissipation of mechanical energy into heat through internal friction between the transported medium layers ($i < 0$).

In integral form, that is as applied to transported medium located between two fixed cross-sections x_1 and x_2, Eq. (1.18) takes the following form

$$\left(\frac{\alpha_{\mathrm{k}}v^2}{2g} + \int\frac{\mathrm{d}p}{\rho g} + z\right)_1 - \left(\frac{\alpha_{\mathrm{k}}v^2}{2g} + \int\frac{\mathrm{d}p}{\rho g} + z\right)_2 = -\int_{x_1}^{x_2} i\,\mathrm{d}x. \tag{1.19}$$

This equation is called the *Bernoulli equation*. It is one of the fundamental equations used to describe the stationary flow of a barotropic medium in a pipeline.

For an *incompressible homogeneous fluid*, which under some conditions can be water, oil and oil product, $\rho = \text{const.}$, $\int \mathrm{d}p/\rho g = p/\rho g + \text{const.}$ Therefore the Bernoulli equation becomes

$$\left(\frac{\alpha_{\mathrm{k}}v^2}{2g} + \frac{p}{\rho g} + z\right)_1 - \left(\frac{\alpha_{\mathrm{k}}v^2}{2g} + \frac{p}{\rho g} + z\right)_2 = -\int_{x_1}^{x_2} i\,\mathrm{d}x.$$

If in addition we take $i = -i_0 = \text{const.}$ ($i_0 > 0$), then

$$\left(\frac{\alpha_{\mathrm{K}}v^2}{2g} + \frac{p}{\rho g} + z\right)_1 - \left(\frac{\alpha_{\mathrm{K}}v^2}{2g} + \frac{p}{\rho g} + z\right)_2 = i_0 \cdot l_{1-2} \tag{1.20}$$

where l_{1-2} is the length of the pipeline between cross-sections 1 and 2.

This last equation has a simple geometric interpretation (see Figure 1.2). This figure illustrates a pipeline profile (heavy broken line); the line $H(x)$ denoting the dependence of the total head H on the coordinate x directed along the axis of the pipeline (straight line) with constant slope β to the horizontal ($i = \mathrm{d}H/\mathrm{d}x = \mathrm{tg}\beta = \text{const.}$) and three components of the total head at an

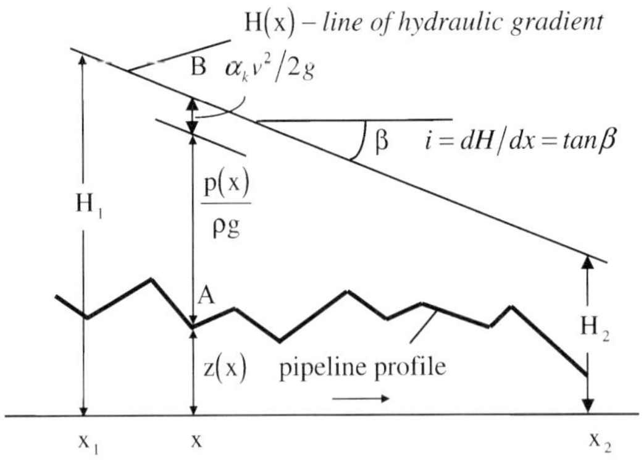

Figure 1.2 Geometric interpretation of the Bernoulli equation.

arbitrary cross-section of the pipeline: geometric head $z(x)$, piezometric head $p(x)/\rho g$ and kinetic head $\alpha_k v^2(x)/2g$.

The line $H(x)$ representing the dependence of the total head H on the coordinate x along the pipeline axis is called the *line of hydraulic gradient.*

It should be noted that if we neglect the dynamic pressure (in oil and oil product pipelines the value of the dynamic pressure does not exceed the pipeline diameter, e.g. at $v \approx 2\ \mathrm{m\,s^{-1}}$, $\alpha_k \approx 1.05$ then $v^2/2g \cong 0.25$ m), and the length of the section between the pipeline profile and the line of hydraulic gradient multiplied by ρg gives the value of the pressure in the pipeline cross-section x. For example, when the length of the section AA (see Figure 1.2) is 500 m and diesel fuel with density $\rho = 840\ \mathrm{kg\,m^{-3}}$ is transported along the pipeline, then

$$\frac{p}{840 \cdot 9.81} = 500 \Rightarrow p = 500 \cdot 840 \cdot 9.81 = 4\,120\,200\ (\mathrm{Pa})$$

or 4.12 MPa (≈42 atm).

1.5.2
Input of External Energy

In fluid flow in the pipeline the mechanical energy is dissipated into heat and the pressure decreases gradually. Devices providing pressure restoration or generation are called *compressors.*

Compressors installed separately or combined in a group form the pumping plant destined to set the fluid moving from the cross-section with lesser pressure to the cross-section with greater pressure. To do this it is required to expend, or deliver from outside to the fluid, energy whose power is denoted by N_{mech}.

Let index 1 in the Bernoulli equation refer to parameters at the cross-section x_1 of the pump entrance (suction line) and index 2 at the cross-section x_2 of the pump exit (discharge line). Since $\rho v S = \text{const.}$, the Bernoulli equation (1.14) may be written as:

$$\int_{x_1}^{x_2} \frac{\mathrm{d}}{\mathrm{d}x}\left[\rho v S \cdot \left(\frac{\alpha_k v^2}{2} + \frac{p}{\rho} + gz\right)\right] \mathrm{d}x = \int_{x_1}^{x_2} n^{\text{in}} \cdot \rho S \, \mathrm{d}x + N_{\text{mech}}.$$

Ignoring the difference between the kinetic and geometric heads we get

$$\rho v S \cdot \frac{p_2 - p_1}{\rho} - \int_{x_1}^{x_2} n^{\text{in}} \cdot \rho S \, \mathrm{d}x = N_{\text{mech}}.$$

Denoting by $\Delta H = (p_2 - p_1)/\rho g$ the *differential head* produced by the pump or pumping plant and taking into account that $\rho v S = \rho Q = \text{const.}$ and $n^{\text{in}} = g v \cdot i$, we obtain

$$N_{\text{mech}} = \rho g Q \cdot \Delta H - \int_{x_1}^{x_2} \rho g Q \cdot i \, \mathrm{d}x = \rho g Q \cdot \Delta H \cdot \left(1 - \int_{x_1}^{x_2} \frac{i}{\Delta H} \mathrm{d}x\right).$$

The expression in parentheses characterizes the loss of mechanical energy within the pump. Usually this factor is taken into account by insertion of the pump efficiency η

$$\eta = \left(1 - \int_{x_1}^{x_2} i/\Delta H \; \mathrm{d}x\right)^{-1} < 1$$

so that

$$N_{\text{mech}} = \frac{\rho g Q \cdot \Delta H}{\eta(Q)}. \tag{1.21}$$

The relation (1.21) is the main formula used to calculate the power of the pump generating head ΔH in fluid pumping with flow rate Q.

1.6 Equation of Change in Internal Motion Kinetic Energy

At the beginning of the previous section it was noted that the total kinetic energy of the transported medium consisted of two terms – the kinetic energy of the center of mass of the particle and the kinetic energy of the internal motion of the center of mass, so that the total energy of a particle is equal to $\alpha_k \rho v^2/2$, where $\alpha_k > 1$. Now we can derive an equation for the second component of the kinetic energy, namely the kinetic energy of the internal or relative motion in the flow of the transported medium.

Multiplication of motion equation (1.10) by the product vS yields

$$\rho S \frac{\mathrm{d}}{\mathrm{d}t}\left(\frac{v^2}{2}\right) = -\frac{\partial p}{\partial x} \cdot v S - \frac{4}{d}\tau_w \cdot v S - \rho g v S \cdot \sin\alpha(x).$$

Subtracting this equation term-by-term from the Bernoulli equation (1.15), one obtains

$$\rho S \frac{\mathrm{d}}{\mathrm{d}t}\left[(\alpha_{\mathrm{k}} - 1)\frac{v^2}{2}\right] = \frac{4}{d}\tau_{\mathrm{w}} \cdot vS + \rho S \cdot n^{\mathrm{in}}.$$

Introduction of $n^{\mathrm{in}} = -gv \cdot i_0$ gives

$$\rho S \frac{\mathrm{d}}{\mathrm{d}t}\left[(\alpha_{\mathrm{K}} - 1)\frac{v^2}{2}\right] = \left(\frac{4}{d}\tau_{\mathrm{w}} \cdot v\right) S - \rho g v S \cdot i_0. \tag{1.22}$$

This is the sought equation of change in *kinetic energy of internal motion* of one-dimensional flow of the transported medium. Its sense is obvious: *the power of the external friction forces* ($4\tau_{\mathrm{w}} \cdot vS/d$) *in one-dimensional flow minus the power* $\rho gS(v \cdot i_0)$ *of internal friction forces between the particles causing transition of mechanical energy into heat is equal to the rate of change of internal motion kinetic energy in the flow of the transported medium.*

For stationary flow ($\mathrm{d}/\mathrm{d}t = 0 + v \cdot \partial/\partial x$) of the transported medium Eq. (1.22) gives

$$\frac{\mathrm{d}}{\mathrm{d}x}\left[(\alpha_{\mathrm{k}} - 1)\frac{v^2}{2}\right] = \frac{4}{d}\frac{\tau_{\mathrm{w}}}{\rho} - g \cdot i_0. \tag{1.23}$$

If $v \cong$ const., which for the flow of an incompressible medium in a pipeline with constant diameter is the exact condition, the left-hand part of the equation vanishes. This means that the tangential friction tension τ_{w} at the pipeline wall and the hydraulic gradient i_0 are connected by

$$\tau_{\mathrm{w}} = \frac{\rho g d}{4} \cdot i_0. \tag{1.24}$$

It must be emphasized that in the general case, including non-stationary flow, such a connection between τ_w and i_0 is absent (see Section 4.1).

1.6.1
Hydraulic Losses (of Mechanical Energy)

The quantity n^{in} entering into Eq. (1.16) denotes the specific power of the internal friction force, that is per unit mass of transported medium. This quantity is very important since it characterizes the loss of mechanical energy converted into heat owing to internal friction between layers of the medium. In order to derive this quantity theoretically one should know how the layers of transported medium move at each cross-section of the pipeline but this is not always possible. In the next chapter it will be shown that in several cases, in particular for laminar, flow such motion can be calculated and the quantity n^{in} can be found. In other cases, such as for turbulent flows of the transported medium, it is not possible to calculate the motion of the layers and other methods of determining n^{in} are needed.

The quantity of specific mechanical energy dissipation n^{in} has the following dimension (from now onwards dimension will be denoted by the symbol [ ])

$$[n^{in}] = \frac{W}{kg} = \frac{J}{s\,kg} = \frac{N\,m}{s\,kg} = \frac{kg\,m\,s^{-2}\,m}{s\,kg} = \frac{m^2}{s^3} = \left[\frac{v^3}{d}\right].$$

So the dimension of n^{in} is the same as the dimension of the quantity v^3/d, hence, without disturbance of generality, one can seek n^{in} in the form

$$n^{in} = -\frac{\lambda}{2} \cdot \frac{v^3}{d} \tag{1.25}$$

where λ is a dimensional factor ($\lambda > 0$), the minus sign shows that $n^{in} < 0$, that is the mechanical energy decreases thanks to the forces of internal friction. The factor 1/2 is introduced for the sake of convenience.

The presented formula does not disturb the generality of the consideration because the unknown dependence of n^{in} on the governing parameters of the flow is accounted for by the factor λ. This dependence is valid for any medium be it fluid, gas or other medium with complex specific properties, e.g. waxy crude oil, suspension or even pulp, that is a mixture of water with large rigid particles.

For stationary fluid or gas flow one can suppose the factor λ to be dependent on four main parameters: the flow velocity v (m s^{-1}), the kinematic viscosity of the flow ν (m^2 s^{-1}), the internal diameter of the pipeline d (m) and the mean height of the roughness of its internal surface Δ (mm or m), so that $\lambda = f(v, \nu, d, \Delta)$. The density of the fluid ρ and the acceleration due to gravity g are not included here because intuition suggests that the friction between fluid or gas layers will be dependent on neither their density nor the force of gravity.

Note that the quantity λ is dimensionless, that is its numerical value is independent of the system of measurement units, while the parameters v, ν, d, Δ are dimensional quantities and their numerical values depend on such a choice. The apparent contradiction is resolved by the well-known Buckingham I-theorem, in accordance with which any dimensionless quantity can depend only on dimensionless combinations of parameters governing this quantity (Lurie, 2001). In our case there are two such parameters

$$\frac{v \cdot d}{\nu} = Re \quad \text{and} \quad \frac{\Delta}{d} = \varepsilon,$$

the first is called the *Reynolds number* and the second the *relative roughness* of the pipeline internal surface. Thus

$$\lambda = \lambda(Re, \varepsilon).$$

The formula (1.25) acquires the form

$$n^{\text{in}} = -\lambda(Re, \varepsilon) \cdot \frac{1}{d} \cdot \frac{v^3}{2}. \tag{1.26}$$

The factor λ in this formula is called the *hydraulic resistance factor*, one of the most important parameters of hydraulics and pipeline transportation. Characteristic values of λ lie in the range 0.01–0.03. More detailed information about this factor and its dependence on the governing parameters will be presented below.

Turning to the hydraulic gradient i_0, one can write

$$i_0 = -\frac{n^{\text{in}}}{gv} = \lambda \cdot \frac{1}{d} \cdot \frac{v^2}{2g}. \tag{1.27}$$

Characteristic values of the hydraulic slope are 0.00005–0.005.

If we substitute Eq. (1.27) into the Bernoulli equation (1.20), we obtain

$$\left(\frac{\alpha_k v^2}{2g} + \frac{p}{\rho g} + z\right)_1 - \left(\frac{\alpha_k v^2}{2g} + \frac{p}{\rho g} + z\right)_2 = \lambda(Re, \varepsilon) \cdot \frac{l_{1-2}}{d} \frac{v^2}{2g}. \tag{1.28}$$

The expression $h_\tau = \lambda \cdot l_{1-2}/d \cdot v^2/2g$ on the right-hand side of this equation is called *the loss of head in Darcy-Veisbach form.*

Using Eq. (1.27) in the case of stationary flow of the transported medium permits us to get an expression for the tangential friction stress τ_w at the pipeline wall. Substitution of Eq. (1.27) into Eq. (1.24), yields

$$\tau_w = \frac{\rho g d}{4} \cdot i_0 = \frac{\rho g d}{4} \cdot \left(\lambda \frac{1}{d} \frac{v^2}{2g}\right) = \frac{\lambda}{4} \cdot \frac{\rho v^2}{2} = C_f \cdot \frac{\rho v^2}{2}, \tag{1.29}$$

$$C_f(Re, \varepsilon) = \frac{\lambda(Re, \varepsilon)}{4}$$

where the dimensional factor C_f is called the *friction factor* of the fluid on the internal surface of the pipeline or the *Funning factor* (Leibenson *et al.*, 1934).

1.6.2
Formulas for Calculation of the Factor $\lambda(Re, \varepsilon)$

Details of methods to find and calculate the factor of hydraulic resistance λ in Eqs. (1.26)–(1.29) and one of the primary factors in hydraulics and pipeline transportation will be given in Chapter 3. Here are shown several formulas exploiting the practice.

If the flow of fluid or gas in the pipeline is laminar, that is jetwise or layerwise (the Reynolds number Re should be less than 2300), then to determine λ the *Stokes formula* (see Section 3.1) is used

$$\lambda = \frac{64}{Re}. \tag{1.30}$$

As the Reynolds number increases ($Re > 2300$) the flow in the pipeline gradually loses hydrodynamic stability and becomes turbulent, that is vortex flow with mixing layers. The best known formula to calculate the factor λ in this case is the *Altshuler formula*:

$$\lambda = 0.11 \cdot \left(\varepsilon + \frac{68}{Re}\right)^{1/4} \tag{1.31}$$

valid over a wide range of Reynolds number from 10^4 up to 10^6 and higher.

If $10^4 < Re < 27/\varepsilon^{1.143}$ and $Re < 10^5$, the Altshuler formula becomes the *Blasius formula*:

$$\lambda = \frac{0.3164}{\sqrt[4]{Re}} \tag{1.32}$$

having the same peculiarity as the Stokes formula for laminar flow, which does not consider the relative roughness of the pipeline internal surface ε. This means that for the considered range of Reynolds numbers the pipeline behaves as a pipeline with a smooth surface. Therefore the fluid flow in this range is *flow in a hydraulic smooth pipe*. In this case the friction tension τ_w at the pipe wall is expressed by formula

$$\tau_w = -\frac{\lambda}{4} \cdot \frac{\rho v^2}{2} = -\frac{0.0791}{\sqrt[4]{vd/\nu}} \cdot \frac{\rho v^2}{2} \approx v^{1.75}$$

signifying that friction resistance is proportional to fluid mean velocity to the power of 1.75.

If $Re > 500/\varepsilon$, the second term in parentheses in the Altshuler formula can be neglected compared to the first one. Whence it follows that at great fluid velocities the fluid friction is caused chiefly by the smoothness of the pipeline internal surface, that is by the parameter ε. In such a case one can use the simpler *Shiphrinson formula* $\lambda = 0.11 \cdot \varepsilon^{0.25}$. Then

$$\tau_w = -\frac{\lambda}{4} \cdot \frac{\rho v^2}{2} = -\frac{0.11 \cdot \varepsilon^{1/4}}{4} \cdot \frac{\rho v^2}{2} \approx v^2.$$

From this it transpires that the friction resistance is proportional to the square of the fluid mean velocity and hence this type of flow is called *square flow*.

Finally, in the region of flow transition from laminar to turbulent, that is in the range of Reynolds number from 2320 up to 10^4 one can use the approximation formula

$$\lambda = \frac{64}{Re} \cdot (1 - \gamma_*) + \frac{0.3164}{\sqrt[4]{Re}} \cdot \gamma_*, \tag{1.33}$$

where $\gamma_* = 1 - e^{-0.002 \cdot (Re - 2320)}$ is the intermittency factor (Ginsburg, 1957). It is obvious that the form of the last formula assures continuous transfer from the Stokes formula for laminar flow to the Blasius formula for turbulent flow in the zone of hydraulic smooth pipes.

To calculate the hydraulic resistance factor λ of the gas flow in a gas main, where the Reynolds number Re is very large and this factor depends only on the condition of the pipeline internal surface, Eq. (1.34) is often used.

$$\lambda = 0.067 \cdot \left(\frac{2\Delta}{d}\right)^{0.2} \tag{1.34}$$

in which the absolute roughness Δ is equal to 0.03–0.05 mm.

Exercise 1. The oil ($\rho = 870$ kg m^{-3}, $\nu = 15$ s St) flows along the pipeline ($D = 156$ mm; $\delta = 5$ mm; $\Delta = 0.1$ mm) with mean velocity $v = 0.2$ m s^{-1}. Determine through the Reynolds criterion the flow regime; calculate factors λ and C_f.

Answer. Laminar; 0.033; 0.0083.

Exercise 2. Benzene ($\rho = 750$ kg m^{-3}, $\nu = 0.7$ s St) flows along the pipeline ($D = 377$ mm; $\delta = 7$ mm; $\Delta = 0.15$ mm) with mean velocity $v = 1.4$ m s^{-1}. Determine through the Reynolds criterion the flow regime; calculate factors λ and C_f.

Answer. Turbulent; 0.017; 0.0041.

Exercise 3. Diesel fuel ($\rho = 840$ kg m^{-3}, $\nu = 6$ s St) flows along the pipeline ($D = 530$ mm; $\delta = 8$ mm; $\Delta = 0.25$ mm) with mean velocity $v = 0.8$ m s^{-1}. Determine the flow regime; calculate factors λ and C_f.

Answer. Turbulent; 0.022; 0.0054.

1.7 Total Energy Balance Equation

Besides the law (1.13) of mechanical energy change of material points, applied to an arbitrary continuum volume in the pipeline there is one more fundamental physical law valid for any continuum – the law of total energy conservation or, as it is also called, the *first law of thermodynamics*. This law asserts that the energy does not appear from anywhere and does not disappear to anywhere. It changes in total quantity from one form into another. As applied to our case this law may be written as follows

$$\frac{d(E_{kin} + E_{in})}{dt} = \frac{dQ^{ex}}{dt} + \frac{dA^{ex}}{dt} \tag{1.35}$$

that is the change in *total energy* ($E_{kin} + E_{in}$) of an arbitrary volume of the transported medium happens only due to the exchange of energy with surrounding bodies owing to external inflow of heat dQ^{ex} and the work of external forces dA^{ex}.

In Eq. (1.35) E_{in} is the internal energy of the considered mass of transported medium, unrelated to the kinetic energy, that is the energy of heat motion, interaction between molecules and atoms and so on. In thermodynamics reasons are given as to why the internal energy is a function of state, that is at thermodynamic equilibrium of a body in some state the energy has a well-defined value regardless of the means (procedure) by which this state was achieved. At the same time the quantities dQ^{ex}/dt and dA^{ex}/dt are not generally derivatives with respect to a certain function of state but only represent the ratio of elementary inflows of heat energy (differential dQ^{ex}) and external mechanical energy (differential dA^{ex}) to the time dt in which these inflows happened. It should be kept in mind that these quantities depend on the process going on in the medium.

In addition to function E_{in} one more function e_{in} is often introduced, representing the internal energy of a unit mass of the considered body $e_{in} = E_{in}/m$, where m is the mass of the body.

We can write Eq. (1.35) for a movable volume of transported medium enclosed between cross-sections $x_1(t)$ and $x_2(t)$. The terms of this equation are

$$\frac{d(E_{kin} + E_{in})}{dt} = \frac{d}{dt}\left[\int_{x_1(t)}^{x_2(t)} \left(\alpha_k \frac{\rho v^2}{2} + \rho e_{in}\right) S\,dx\right],$$

$$\frac{dQ^{ex}}{dt} = \int_{x_1(t)}^{x_2(t)} \pi d \cdot q_n\,dx,$$

$$\frac{dA^{ex}}{dt} = -\int_{x_1(t)}^{x_2(t)} \frac{\partial}{\partial x}(pSv)\,dx - \int_{x_1(t)}^{x_2(t)} \rho g \sin\alpha \cdot v \cdot S\,dx + N_{mech}$$

where q_n is the heat flux going through the unit area of the pipeline surface per unit time (W m^{-2}); $\pi d \cdot dx$ is an element of pipeline surface area and d is the pipeline diameter.

Gathering all terms, we obtain

$$\frac{d}{dt}\left[\int_{x_1(t)}^{x_2(t)} \left(\alpha_k \cdot \frac{\rho v^2}{2} + \rho e_{in}\right) S\,dx\right] = \int_{x_1(t)}^{x_2(t)} \pi d \cdot q_n\,dx$$
$$- \int_{x_1(t)}^{x_2(t)} \frac{\partial}{\partial x}(pSv)\,dx - \int_{x_1(t)}^{x_2(t)} \rho g \sin\alpha \cdot v \cdot S\,dx + N_{mech}.$$

Differentiation of the left-hand side of this equation gives

$$\int_{x_1(t)}^{x_2(t)} \left\{\frac{\partial}{\partial t}\left[\left(\frac{\alpha_k v^2}{2} + e_{in}\right)\rho S\right] + \frac{\partial}{\partial x}\left[\left(\frac{\alpha_k v^2}{2} + e_{in}\right)\rho v S\right]\right\} dx$$
$$= \int_{x_1(t)}^{x_2(t)} \pi d \cdot q_n\,dx - \int_{x_1(t)}^{x_2(t)} \frac{\partial}{\partial x}\left(\frac{p}{\rho}\rho v S\right) dx$$
$$- \int_{x_1(t)}^{x_2(t)} \rho v S g \frac{\partial z}{\partial x}\,dx + N_{mech}$$

or

$$\int_{x_1(t)}^{x_2(t)} \left\{ \frac{\partial}{\partial t} \left[\left(\frac{\alpha_k v^2}{2} + e_{in} \right) \rho S \right] + \frac{\partial}{\partial x} \left[\left(\frac{\alpha_k v^2}{2} + e_{in} + \frac{p}{\rho} \right) \rho v S \right] \right\} dx$$
$$= \int_{x_1(t)}^{x_2(t)} \pi d \cdot q_n \, dx - \int_{x_1(t)}^{x_2(t)} \rho v S g \frac{\partial z}{\partial x} \, dx + N_{mech}. \quad (1.36)$$

If we assume that inside the region $[x_1(t), x_2(t)]$ the external sources of mechanical energy are absent, that is $N_{mech} = 0$, then it is possible to pass from integral equality (1.36) to the corresponding differential equation using, as before, the condition that this equation should be true for any volume of the transported medium, that is the limits of integration $x_1(t)$ and $x_2(t)$ in (1.36) are to be arbitrarily chosen. Then the sign of the integral can be omitted and the differential equation is

$$\frac{\partial}{\partial t} \left[\left(\frac{\alpha_k v^2}{2} + e_{in} \right) \rho S \right] + \frac{\partial}{\partial x} \left[\left(\frac{\alpha_k v^2}{2} + e_{in} + \frac{p}{\rho} \right) \rho v S \right]$$
$$= \pi d \cdot q_n - \rho v S g \frac{\partial z}{\partial x}. \quad (1.37)$$

Excluding from Eq. (1.37) the change in kinetic energy with the help of the Bernoulli equation with term by term subtraction of Eq. (1.16) from Eq. (1.37) we get one more energy equation

$$\rho S \frac{\partial}{\partial t} \left(\frac{\alpha_k v^2}{2} \right) + \rho v S \frac{\partial}{\partial x} \left(\frac{\alpha_k v^2}{2} + \int \frac{dp}{\rho} + gz \right) = \rho v S g \cdot i$$

called the *equation of heat inflow*.

This equation could be variously written. First, it may be written through the internal energy e_{in}:

$$\frac{\partial}{\partial t}(e_{in} \cdot \rho S) + \frac{\partial}{\partial x}(e_{in} \cdot \rho v S) = \pi d \cdot q_n - p \frac{\partial v S}{\partial x} - \rho v S g \cdot i$$

or

$$\rho S \left(\frac{\partial e_{in}}{\partial t} + v \frac{\partial e_{in}}{\partial x} \right) = \pi d \cdot q_n - p \cdot \frac{\partial v S}{\partial x} - \rho v S g \cdot i. \quad (1.38)$$

This equation proved to be especially convenient for modeling flows of incompressible or slightly compressible fluids because the derivative $\partial(vS)/\partial x$ expressing the change in fluid volume in the pipeline cross-section is extremely small as is the work $p \cdot \partial(vS)/\partial x$ of the pressure forces. With this in mind Eq. (1.38) may be written in a particularly simple form:

$$\rho \frac{de_{in}}{dt} \cong \frac{4}{d} \cdot q_n - \rho v g \cdot i. \quad (1.39)$$

This means that the rate of internal energy change of the transported medium is determined by the inflow of external heat through the pipeline surface and heat extraction due to conversion of mechanical energy into heat produced by friction between the continuum layers.

Second, the equation of heat inflow can be written using the function $J = e_{in} + p/\rho$ representing one of the basic thermodynamic functions, *enthalpy or heat content,* of the transported medium

$$\frac{\partial}{\partial t}(e_{in} \cdot \rho S) + \frac{\partial}{\partial x}\left[\left(e_{in} + \frac{p}{\rho}\right)\rho v S\right] = \pi d \cdot q_n + \rho v S g \cdot \left(\frac{1}{\rho g}\frac{\partial p}{\partial x} - i\right)$$

or

$$\frac{\partial}{\partial t}(e_{in} \cdot \rho S) + \frac{\partial}{\partial x}[J \cdot \rho v S] = \pi d \cdot q_n + \rho v S g \cdot \left(\frac{1}{\rho g}\frac{\partial p}{\partial x} - i\right). \tag{1.40}$$

If we take into account (as will be shown later) that the expression in parentheses on the right-hand side of this equation is close to zero, since for a relatively light medium, e.g. gas, the hydraulic slope is expressed through the pressure gradient by the formula $i = 1/\rho g \cdot \partial p/\partial x$, the equation of heat inflow can be reduced to a simpler form

$$\frac{\partial \rho S \cdot e_{in}}{\partial t} + \frac{\partial \rho v S \cdot J}{\partial x} \cong \pi d \cdot q_n \tag{1.41}$$

in which the dissipation of mechanical energy appears to be absent.

Temperature Distribution in Stationary Flow

The equation of heat inflow in the form (1.39) or (1.41) is convenient to determine the temperature distribution along the pipeline length in stationary flow of the transported medium.

1. For an *incompressible* or *slightly compressible* medium, e.g. dropping liquid: water, oil and oil product, this equation has the form

$$\rho v \cdot \frac{de_{in}}{dx} \cong \frac{4}{d} \cdot q_n - \rho v g \cdot i. \tag{1.42}$$

The internal energy e_{in} depends primarily on the temperature of the fluid T, the derivative de_{in}/dT giving its specific heat C_v (J kg^{-1} K^{-1}). If we take $C_v = \text{const.}$ then $e_{in} = C_v \cdot T + \text{const.}$

To model the heat flux q_n the *Newton formula* is usually used

$$q_n = -\kappa \cdot (T - T_{ex}), \tag{1.43}$$

by which this flow is proportional to the difference between the temperatures T and T_{ex} in and outside the pipeline, with $q_n < 0$ when $T > T_{ex}$ and $q_n > 0$ when $T < T_{ex}$. The factor κ (W m^{-2} K^{-1}) in this formula characterizes the overall heat resistance of the materials through which the heat is transferred from the pipe to the surrounding medium (anticorrosive and heat insulation, ground, the boundary between ground and air and so on) or the reverse. This factor is called the *heat-transfer factor*.

The hydraulic gradient i can sometimes be considered constant $i = -i_0 \approx$ const., if the dissipation of mechanical energy in the stationary fluid flow in the pipeline with constant diameter is identical at all cross-sections of the pipeline.

With due regard for all the aforesaid Eq. (1.42) is reduced to the following ordinary differential equation

$$\rho C_v v \cdot \frac{dT}{dx} = -\frac{4\kappa}{d}(T - T_{ex}) + \rho v g i_0 \tag{1.44}$$

for temperature $T = T(x)$. From this equation in particular it follows that the heat transfer through the pipeline wall (the first term on the right-hand side) lowers the temperature of the transported medium when $T(x) > T_{ex}$ or raises it when $T(x) < T_{ex}$, whereas the dissipation of mechanical energy (the second term on the right-hand side) always implies an increase in the temperature of the transported medium.

The solution of the differential equation (1.44) with initial condition $T(0) = T_0$ yields

$$\frac{T(x) - T_{ex} - T_{\otimes}}{T_0 - T_{ex} - T_{\otimes}} = \exp\left(-\frac{\pi d\kappa}{C_v \dot{M}} x\right). \tag{1.45}$$

Where $T_{\otimes} = g i_0 \dot{M}/\pi d\kappa$ is a constant having the dimension of temperature; $\dot{M} = \rho v S$ is the mass flow rate of the fluid ($\dot{M} =$ const.). The formula thus obtained is called the *Shuchov formula*.

Figure 1.3 illustrates the distribution of temperature $T(x)$ along the pipeline length x in accordance with Eq. (1.45).

The figure shows that when the initial temperature T_0 is greater than $(T_{ex} + T_{\otimes})$, the moving medium cools down, while when T_0 is less than $(T_{ex} + T_{\otimes})$, the medium gradually heats up. In all cases with increase in the pipeline length the temperature $T \to (T_{ex} + T_{\otimes})$.

In particular from Eq. (1.44) it follows that if the heat insulation of the pipeline is chosen such that at the initial cross-section of the pipeline $x = 0$

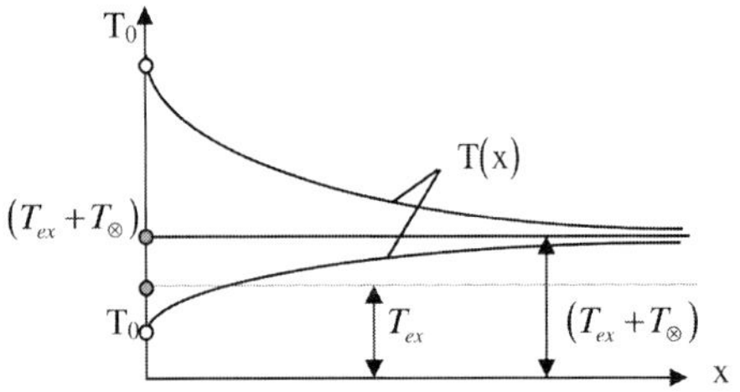

Figure 1.3 Temperature distribution along the pipeline length.

the condition of equality to zero of the right-hand side is obeyed

$$-\frac{4\kappa}{d}(T_0 - T_{ex}) + \rho v g i_0 = 0$$

that is the factor κ satisfies the condition

$$\kappa = \frac{\rho v d g i_0}{4(T_0 - T_{ex})} = \frac{g\dot{M} \cdot i_0}{\pi d \cdot (T_0 - T_{ex})}.$$

And the temperature of the transported medium would remain constant and equal to its initial value over the whole pipeline section. In such a case the heat outgoing from the pipeline would be compensated by the heat extracted by internal friction between the layers. Such an effect is used, for example, in oil transportation along the Trans-Alaska oil pipeline (USA, see the cover picture). Through good insulation of the pipeline the oil is pumped over without preheating despite the fact that in winter the temperature of the environment is very low.

From Eq. (1.45) follows the connection between the initial T_0 and final T_L temperatures of the transported medium. If in this formula we set $x = L$, where L is the length of the pipeline section, we obtain

$$\frac{T_L - T_{ex} - T_\otimes}{T_0 - T_{ex} - T_\otimes} = \exp\left(-\frac{\pi d \kappa L}{C_v \dot{M}}\right). \tag{1.46}$$

Expressing now from (1.46) the argument under the exponent and substituting the result in Eq. (1.45), we get the expression for the temperature distribution through the initial and final values

$$\frac{T(x) - T_{ex} - T_\otimes}{T_0 - T_{ex} - T_\otimes} = \left(\frac{T_L - T_{ex} - T_\otimes}{T_0 - T_{ex} - T_\otimes}\right)^{x/L}. \tag{1.47}$$

Exercise 1. The initial temperature of crude oil ($\rho = 870\ \mathrm{kg\ m^{-3}}$, $C_v = 2000\ \mathrm{J\ kg^{-1}\ K^{-1}}$, $Q = 2500\ \mathrm{m^3\ h^{-1}}$), pumping over a pipeline section ($d = 800$ mm, $L = 120$ km, $i_0 = 0.002$) is 55 °C. The temperature of the surrounding medium is 8 °C. The heat insulation of the pipeline is characterized by the heat-transfer factor $\kappa = 2\ \mathrm{W\ m^{-2}\ K^{-1}}$. It is required to find the temperature at the end of the section.

Solution. Calculate first the temperature $T_\otimes$:

$$T_\otimes = \frac{g i_0 \dot{M}}{\pi d \kappa} = \frac{9.81 \cdot 0.002 \cdot 870 \cdot (2500/3600)}{3.14 \cdot 0.8 \cdot 2} \cong 2.36\ \mathrm{K}.$$

Using Eq. (1.46) we obtain

$$\frac{T_L - 8 - 2.36}{55 - 8 - 2.36} = \exp\left(-\frac{3.14 \cdot 0.8 \cdot 2 \cdot 120 \cdot 10^3}{2000 \cdot 870 \cdot (2500/3600)}\right),$$

from which follows $T_L \cong 37.5$ °C.

Exercise 2. By how much would the temperature of the oil ($C_v = 1950\ \mathrm{J\,kg^{-1}\,K^{-1}}$) be raised due to the heat of internal friction when the oil is transported by an oil pipeline ($L = 150$ km, $d = 500$ mm, $i_0 = 0.004$) provided with ideal heat insulation ($\kappa = 0$)?

Solution. In this case it is impossible to use at once Eq. (1.45) since $\kappa = 0$. To use Eq. (1.47) one should go to the limit at $\kappa \to 0$, therefore it would be better to use Eq. (1.44)

$$\rho C_v v \cdot \frac{dT}{dx} = \rho v g i_0 \quad \text{or} \quad C_v \cdot \frac{dT}{dx} = g i_0,$$

from which $\Delta T = g i_0 L / C_v = 9.81 \cdot 0.004 \cdot 150 \cdot 10^3 / 1950 \cong 3$ K.

Exercise 3. It is required to obtain the temperature of oil pumping over the pipeline section of length 150 km in cross-sections $x = 50$, 100 and 125 km, if the temperature at the beginning of the pipeline $T_0 = 60\,^\circ$C, that at the end $T_L = 30\,^\circ$C, and that of the environment $T_{ex} = 8\,^\circ$C. The extracted heat of internal friction may be ignored.

Solution. Using Eq. (1.46), one gets

$$\frac{T(x) - 8}{60 - 8} = \left(\frac{30 - 8}{60 - 8}\right)^{x/L} \quad \text{and} \quad T(x) = 8 + 52 \cdot (0.4231)^{x/150}.$$

Substitution in this formula of successive $x = 50$, 100 and 125 gives $T(50) \cong 47\,^\circ$C; $T(100) \cong 37.3\,^\circ$C; $T(125) \cong 33.4\,^\circ$C.

2. For stationary flow of a *compressible medium*, e.g. gas, the equation of heat inflow (1.41) takes the form

$$\rho v S \frac{dJ}{dx} = \pi d \cdot q_n.$$

In the general case, the gas enthalpy J is a function of pressure and temperature $J = J(p, T)$, but for a *perfect gas*, that is a gas obeying the Clapeyron law $p = \rho R T$, where R is the gas constant, the enthalpy is a function only of temperature $J = C_p \cdot T + \text{const.}$, where C_p is the gas specific heat capacity at constant pressure ($C_p > C_v$; $C_p - C_v = R$). Regarding $C_p = \text{const.}$ and taking as before $q_n = -\kappa \cdot (T - T_{ex})$, we transform the last equation to

$$C_p \dot{M} \frac{dT}{dx} = -\pi d \kappa \cdot (T - T_{ex})$$

or

$$\frac{dT}{dx} = -\frac{\pi d \kappa}{C_p \dot{M}} \cdot (T - T_{ex}).$$

The solution of this differential equation with initial condition $T(0) = T_0$ gives

$$\frac{T(x) - T_{ex}}{T_0 - T_{ex}} = \exp\left(-\frac{\pi d \kappa}{C_p \dot{M}} x\right), \tag{1.48}$$

which is similar to the solution (1.45) for temperature distribution in an incompressible fluid. The difference consists only in that instead of heat capacity C_v in the solution (1.47) we use heat capacity C_p and the temperature $T_\otimes$ taking into account the heat of internal friction is absent (for methane $C_p \cong 2230\ \mathrm{J\ kg^{-1}\ K^{-1}}$; $C_v \cong 1700\ \mathrm{J\ kg^{-1}\ K^{-1}}$).

The temperature T_L of the gas at the end of the gas pipeline section is found from

$$\frac{T_L - T_{ex}}{T_0 - T_{ex}} = \exp\left(-\frac{\pi d \kappa L}{C_p \dot{M}}\right) \tag{1.49}$$

with regard to which the distribution (1.47) takes the form

$$\frac{T(x) - T_{ex}}{T_0 - T_{ex}} = \left(\frac{T_L - T_{ex}}{T_0 - T}\right)^{x/L} \tag{1.50}$$

allowing us to express the temperature through the initial and final temperatures.

Note that for a real gas the enthalpy $J = J(p, T)$ of the medium depends not only on temperature but also on pressure, so the equation of heat inflow has a more complex form. By the dependence $J(p, T)$ is explained, in particular, the *Joule-Thomson effect*.

1.8
Complete System of Equations for Mathematical Modeling of One-Dimensional Flows in Pipelines

This system consists of the following equations.

1. Continuity equation (1.6)

$$\frac{\partial \rho S}{\partial t} + \frac{\partial \rho v S}{\partial x} = 0;$$

2. Momentum (motion) equation (1.10)

$$\rho\left(\frac{\partial v}{\partial t} + v\frac{\partial v}{\partial x}\right) = -\frac{\partial p}{\partial x} - \frac{4}{d}\tau_w - \rho g \sin\alpha(x);$$

3. Equation of mechanical energy balance (1.15)

$$\frac{\partial}{\partial t}\left(\frac{\alpha_k v^2}{2}\right) + v \cdot \frac{\partial}{\partial x}\left(\frac{\alpha_k v^2}{2} + P(\rho) + gz\right) = vg \cdot i;$$

4. Equation of total energy balance (1.37)

$$\frac{\partial}{\partial t}\left[\left(\frac{\alpha_k v^2}{2}+e_{in}\right)\rho S\right]+\frac{\partial}{\partial x}\left[\left(\frac{\alpha_k v^2}{2}+J\right)\rho v S\right] = \pi d\cdot q_n - \rho v g S\frac{dz}{dx}.$$

The number of unknown functions in this equation is 10: ρ, v, p, S, e_{in}, T, τ_w, i, q_n, α_k, while the number of equations is 4. Therefore there are needed additional relations to *close* the system of equations. As *closing relations* the following relations are commonly used:

- equation of state $p = p(\rho, T)$, characterizing the properties of the transported medium;
- equation of pipeline state $S = S(p, T)$ characterizing the deformation ability of the pipeline;
- calorimetric dependences $e_{in} = e(p, T)$ or $J = J(p, T)$;
- dependence $q_n = -\kappa \cdot (T - T_{ex})$ or more complex dependences representing heat exchange between the transported medium and the environment;
- hydraulic dependence $\tau_w = \tau_w(\rho, v, \dot{v}, d, \nu, \ldots)$;
- dependences $\alpha_k = f(\rho, v, \nu, d, \ldots)$, or $i = \tilde{f}(\tau_w)$,

characterizing internal structure of medium flow.

To obtain closing relations a more detailed analysis of flow processes is needed. It is also necessary to consider mathematical relations describing properties of the transported medium and the pipeline in which the medium flows.

The division of mechanics in which properties of a transported medium such as viscosity, elasticity, plasticity and other more complex properties are studied is called *rheology*.

2
Models of Transported Media

Algebraic relations connecting parameters of the transported medium such as density, pressure, temperature and so on are called *equations of state*. Each of these relations represents of course a certain schematization of the properties of the considered medium and is only a *model* of a given medium. Let us consider some models.

2.1
Model of a Fluid

By *fluid* is meant a continuum in which the interaction of the contacting interior parts at rest is reduced only to the pressing force of pressure. If fluid particles interact along the surface element $d\sigma$ with the unit normal $\boldsymbol{n}$ (Figure 2.1), the force $d\boldsymbol{F}_n$ with which the fluid particles on one side of the element act on the fluid particles on the other side of the element is proportional to the area $d\sigma$, is directed along the normal $\boldsymbol{n}$ and has a pressing action on them. Then

$$d\boldsymbol{F}_n = -p\boldsymbol{n}\, d\sigma. \tag{2.1}$$

The magnitude p of this force does not depend on the surface element orientation and is called *pressure*.

Thus, $p = |\, d\boldsymbol{F}_n / d\sigma|$. The absence of tangential friction forces in the state of rest models the fact that the fluid takes the shape of the vessel it fills.

Further classification of fluids is dependent on whether or not tangential friction forces are taken into account on exposure to fluid flow. In accordance

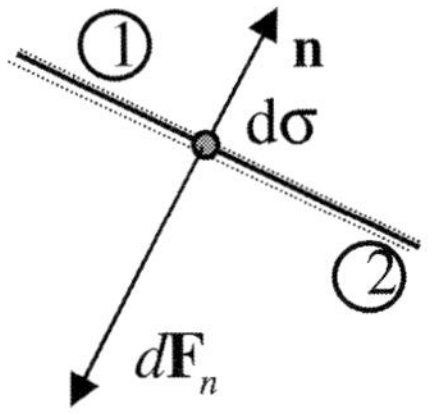

Figure 2.1 A scheme of force interactions in a fluid.

Modeling of Oil Product and Gas Pipeline Transportation. Michael V. Lurie

ISBN: 978-3-527-40833-7

with this there are two models: the model of an *ideal fluid* and the model of a *viscous fluid*.

2.2
Models of Ideal and Viscous Fluids

In the model of an *ideal fluid* it is assumed that tangential friction forces between fluid particles separated by an elementary surface are absent, not only in the state of rest but also in the state of flow. Such a schematization (or model) of a fluid appears to be very fruitful when the tangential components of interaction forces, that is friction forces, are far smaller than their normal components, that is pressure forces. In other cases when the friction forces are comparable with or even exceed the pressure forces the model of an ideal fluid has proved to be inapplicable. Hence expression (2.1) for an ideal fluid is true in the state of rest as well as in the state of flow.

In the *model of a viscous fluid* tangential stresses resulting in fluid flow are taken into account. Let for example the fluid layers move as shown in Figure 2.2.

Here $u(y)$ is the velocity distribution in the flow and y is the direction of a normal to the elementary surface $d\sigma$.

In the model of a viscous fluid it is accepted that the tangential stress τ between the layers of the moving fluid is proportional to the velocity difference of these layers calculated per unit length of the distance between them, namely to the velocity gradient du/dy:

$$\tau = \mu \frac{du}{dy}. \tag{2.2}$$

The tangential stress τ is defined as the friction force between the fluid layers divided by the area of the surface separating these layers. Then the dimension of the stress τ is

$$[\tau] = \frac{\text{force}}{\text{area}} = \frac{M \times L/T^2}{L^2} = \frac{M}{L \times T^2}.$$

In the SI system of units the stress τ is measured by $\text{Pa} = \text{kg m}^{-1}\,\text{s}^{-2}$.

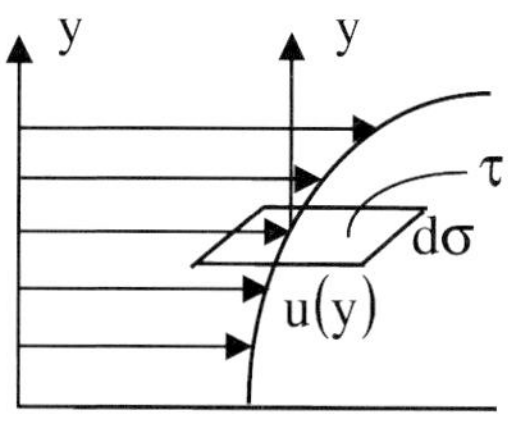

Figure 2.2 Illustration of the definition of the viscous friction law.

The proportionality factor μ in the law of viscous friction (2.2) is called the *factor of dynamic viscosity* or simply *dynamic viscosity*. Its dimension is

$$[\mu] = [\tau] \times \mathrm{T} = \frac{\mathrm{M}}{\mathrm{L} \times \mathrm{T}}$$

In SI units μ is measured in kg $\mathrm{m}^{-1}\mathrm{s}^{-1}$ and is expressed through poise P, where 1 P = 0.1 kg m^{-1} s^{-1}. For example the dynamic viscosity of water is equal to 0.01 P = 0.001 kg m^{-1} s^{-1} = 1 cP (centipoise).

The factor of kinematic viscosity or simply *kinematic viscosity* ν of a fluid is defined as the ratio μ/ρ, therefore

$$[\nu] = \left[\frac{\mu}{\rho}\right] = \frac{\mathrm{M}/(\mathrm{L} \times \mathrm{T})}{\mathrm{M}/\mathrm{L}^3} = \frac{\mathrm{L}^2}{\mathrm{T}}.$$

In SI units ν is measured in $\mathrm{m}^2\ \mathrm{s}^{-1}$ and expressed in stokes St, where 1 St = 10^{-4} $\mathrm{m}^2\ \mathrm{s}^{-1}$. For example, the kinematic viscosity of water is equal to 0.01 St = 10^{-6} $\mathrm{m}^2\ \mathrm{s}^{-1}$ = 1 cSt (centistoke). The kinematic viscosity of benzene is approximately equal to 0.6 cSt; that of diesel fuel is 4–9 cSt and that of low-viscous oil 5–15 cSt.

The viscosity of oil and of almost all oil products depends on temperature. As the temperature increases the viscosity decreases, whereas reduction in temperature leads to viscosity enhancement. To calculate the dependence of the kinematic viscosity ν on temperature T it different formulas can be used including the *Reynolds–Filonov formula*

$$\nu(T) = \nu_0 \cdot \mathrm{e}^{-\kappa\cdot(T-T_0)} \tag{2.3}$$

in which ν_0 is the kinematic viscosity of a fluid at temperature T_0 and κ (1/ K) is an empirical factor. Equation (2.3) means that the fluid viscosity varies exponentially with temperature.

To use Eq. (2.3) it is necessary to know the factor κ or the viscosity ν_1 of the same fluid at another temperature T_1. Then this factor is found from the relation

$$\kappa = \frac{\ln(\nu_0/\nu_1)}{(T_1 - T_0)}. \tag{2.4}$$

Exercise. The kinematic viscosity of summer diesel fuel at temperature +20 °C is 5 cSt, whereas at a temperature of 0 °C it increases to 8 cSt. Determine the viscosity of the same fuel at temperature +10 °C.

Solution. With Eq. (2.4) the factor $\kappa = \ln(5/8)/(0 - 20) \cong 0.0235$ is determined. The viscosity is found from Eq. (2.3) $\nu = 5 \cdot \exp[-0.0235 \cdot (10 - 20)] \cong$ 6.3 cSt.

2.3
Model of an Incompressible Fluid

A fluid is called *incompressible* if its density does not vary when moving, that is $d\rho/dt = 0$. If initially all fluid particles had equal densities, that is the fluid was *homogeneous*, it remains homogeneous as before, namely $\rho = \rho_0 = \text{const.}$ while moving.

Of course an incompressible fluid is only a model of a real medium, since as is known absolutely incompressible media do not exist, but when the change in fluid density in a certain process can be neglected the model of an incompressible fluid may be very useful. For example, under normal conditions water density is 1000 kg m^{-3}, benzene $\approx$ 735–750 kg m^{-3}, diesel fuel $\approx$ 840 kg m^{-3}, oil $\approx$ 870–900 kg m^{-3} etc.

If the fluid density depends only on pressure, that is $\rho = \rho(p)$, the condition of incompressibility is $d\rho/dt = 0$ being tantamount to $d\rho/dp = 0$.

2.4
Model of Elastic (Slightly Compressible) Fluid

There are processes which require that account is taken of even a small variation in fluid density. In such a case the so-called *elastic fluid* model is often used. In this model the fluid density depends on pressure as follows

$$\rho(p) = \rho_0[1 + \beta(p - p_0)] \tag{2.5}$$

where β (1/Pa) is the *compressibility factor*; ρ_0 the fluid density at normal pressure p_0. The compressibility factor is the inverse of the *elastic modulus* K (Pa), that is $K = 1/\beta$. Then Eq. (2.5) reduces to

$$\rho(p) = \rho_0\left(1 + \frac{p - p_0}{K}\right). \tag{2.6}$$

Mean values of the elastic modulus for oil and oil products vary in the range 1400–1500 MPa, so that $K \approx 1.4 - 1.5 \cdot 10^9$ Pa. It follows that the deviation of density ρ from the normal $\rho_0 : \Delta\rho = \rho_0 \cdot (p - p_0)/K$ is very small for oil and oil products. For example for a fluid with $\rho_0 = 850$ kg m^{-3} at $p - p_0 = 5$ MPa ($\approx$ 50 atm) the deviation $\Delta\rho$ is 2.8 kg m^{-3}.

2.5
Model of a Fluid with Heat Expansion

The expansion of different media on heating and subsequent compression on cooling is taken into account in the fluid model with volume expansion. In the

Table 2.1 Factor of volume expansion ξ.

Density ρ, kg m^{-3}	Factor ξ, K^{-1}
720–739	0.001183
740–759	0.001118
760–779	0.001054
780–799	0.000995
800–819	0.000937
820–839	0.000882
840–859	0.000831
860–880	0.000782

model to be considered the density ρ is a function of temperature T, so that $\rho = \rho(T)$

$$\rho(T) = \rho_0[1 + \xi(T_0 - T)] \tag{2.7}$$

in which ξ (1/K) is the factor of volume expansion, ρ_0 and T_0 are the density and temperature of the fluid under normal conditions (often $T_0 = 293$ K (20 °C); $\rho_0 = \rho(p_0, T_0)$; $p_0 = p_{st} = 101325$ Pa). The values of the factor ξ for oil and oil products are given in Table 2.1.

From Eq. (2.6) it follows that on heating, that is at $T > T_0$, the fluid expands, that is $\rho < \rho_0$ whereas at $T < T_0$, $\rho > \rho_0$, that is the fluid is compressed.

Exercise. The density ρ_0 of benzene at 20 °C is 745 kg m^{-3}. Determine the density of the same benzene at 10 °C.

Solution. Using Eq. (2.6) and Table 2.1 we have

$\rho(10\,°\text{C}) = 745 \cdot [1 + 0.00118 \cdot (20 - 10)] = 753.3$ kg m^{-3}.

Thus the density is increased by 8.3 kg m^{-3}.

There is also used a model for fluid expansion with regard to baric and heat expansion. In such a model the density is a function of pressure and temperature $\rho = \rho(p, T)$ called the state equation

$$\rho(p, T) = \rho_0 \left[1 + \xi(T - T_0) + \frac{p - p_0}{K}\right]. \tag{2.8}$$

Here p_0, T_0 are the nominal pressure and temperature satisfying the relation $\rho_0 = \rho(p_0, T_0)$.

Exercise. The density of benzene ρ_0 at 20 °C and atmospheric pressure $p_{at} \approx 0.1$ MPa is 745 kg m^{-3}. Determine the density of the same benzene at temperature 10 °C and pressure 6.5 MPa.

Solution. Using Eq. (2.7) and Table 2.1 we get $\rho(p, T) = 745 \cdot [1 + 0.00118 \cdot (20 - 10) + (6.5 - 0.1) \cdot 10^6/(1.5 \cdot 10^9)] = 757$ kg m^{-3}. The density is increased by 12 kg m^{-3}.

2.6 Models of Non-Newtonian Fluids

Fluids modeled by condition (2.2) of viscous friction are called *Newtonian viscous fluids* in accordance with the name of the law (2.2). The quantity $\dot{\gamma} = du/dy$ having the sense of velocity gradient with dimension s^{-1}, is called the *shear rate*. The linear dependence (2.2) between the tangential friction stress τ and the shear rate $\dot{\gamma}$ is shown in Figure 2.3. This dependence states: "*Tangential stresses arising in a medium having been modeled by a Newtonian viscous fluid are proportional to the shear rate of the fluid layers relative to each other. When the shear rate vanishes, the tangential friction stresses also disappear.*"

The dynamic viscosity of fluid μ is represented in this model by the slope of a straight line on the plane $(\dot{\gamma}, \tau) : \mu = \tan\varphi$, where φ is the angle of inclination of the straight line to the abscissa. Many experiments have shown that the model of a Newtonian viscous fluid well schematizes processes taking place in many real fluids.

And yet there exist dependences of τ on $\dot{\gamma}$ (*flow curve*) that differ significantly from that depicted in Figure 2.3. Such fluids are called *non-Newtonian*.

As an example of a non-Newtonian fluid is a model of a *power Ostwald fluid* (Wilkinson, 1960)

$$\tau = k \cdot \left|\frac{du}{dy}\right|^{n-1} \cdot \frac{du}{dy} \tag{2.9}$$

in which the relation of the tangential friction stresses between fluid layers has a power nature. In other words the apparent viscosity $\tilde{\mu}$ of such a fluid does not remain constant as in the model of Newtonian fluid, but depends on the characteristics of the flow, namely

$$\tilde{\mu} = k \cdot \left|\frac{du}{dy}\right|^{n-1}. \tag{2.10}$$

In this model k and n are factors. Fluids with $n < 1$ are called *pseudo-plastic fluids*. These models are applied to describe the behavior of suspension flows,

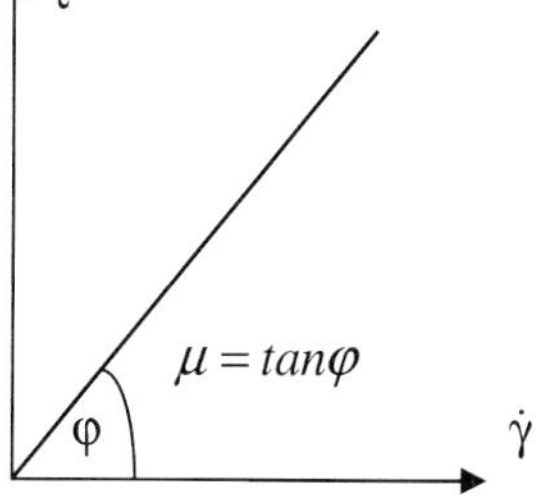

Figure 2.3 A model of a Newtonian viscous fluid.

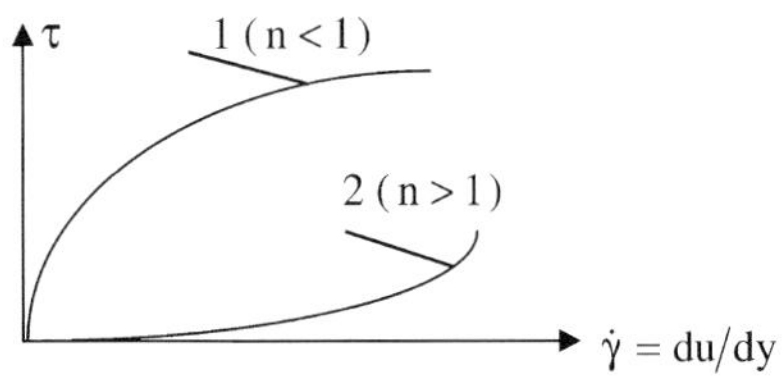

Figure 2.4 Models of non-Newtonian fluids: 1, pseudo-plastic fluids; 2, dilatant fluids.

that is viscous fluids with suspended small particles. Flow curves of such fluids have the form of curve 1 in Figure 2.4 (Wilkenson, 1960).

Fluids with $n > 1$ are called *dilatant fluids*. Starch glue is an example of fluid whose behavior is described by the dilatant model. Flow curves of these fluids have the form of curve 2 in Figure 2.4 (Wilkenson, 1960).

Model of viscous-plastic medium with limit shear stress; model of Shvedov–Bingham fluid (Wilkenson, 1960). There are fluids in which the stresses between the fluid layers are sufficiently well described by the following relations

$$\tau = \tau_0 + \mu\frac{du}{dy}, \text{ at } \tau > \tau_0;$$
$$\dot{\gamma} = \frac{du}{dy} = 0, \text{ at } |\tau| \leq \tau_0; \tag{2.11}$$
$$\tau = -\tau_0 + \mu\frac{du}{dy}, \text{ at } \tau < -\tau_0.$$

These expressions imply that the flow of such a fluid does not begin as long as the absolute value of the tangential friction stress τ does not exceed a limiting value τ_0, being the characteristic of the given medium and called the *limit shear stress*. In this case $\dot{\gamma} = 0$. At $|\tau| \geq \tau_0$ and $\dot{\gamma} \neq 0$ the medium flows as a viscous fluid.

Real media whose properties are satisfactory modeled by the viscous-plastic Bingham fluid are, for example, high-paraffinaceous and solidifying oils, mud solutions, lacquers, paints and other media.

Flow curves of viscous-plastic media are shown in Figure 2.5.

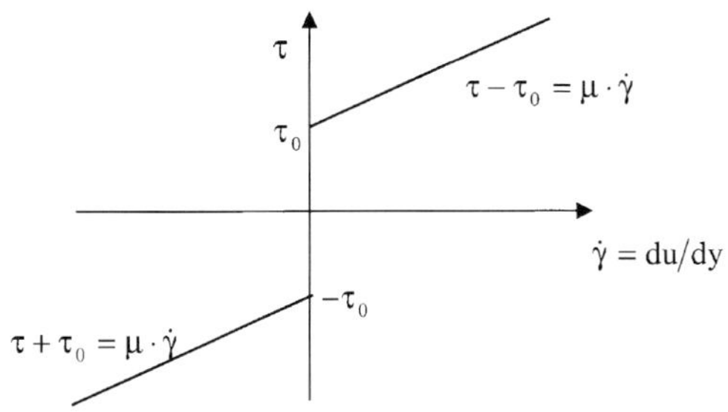

Figure 2.5 Flow curves of a viscous-plastic fluid.

2.7 Models of a Gaseous Continuum

We proceed now to the description of the basic models used for gas flow. First let us consider the properties common to all gases. One such property is that for all gases in a state of thermodynamic equilibrium there is a relation between pressure p, absolute temperature T and density ρ (or specific volume $\upsilon = 1/\rho$)

$$\Phi(p, \upsilon, T) = 0 \tag{2.12}$$

called the *equation of state*. The physical nature of this fact is discussed in text-books on statistical physics and thermodynamics. In most models it is also assumed that when the motion starts the relation (2.12) remains. This means that the establishment of thermodynamic equilibrium happens much faster than the non-equilibrium caused by the resulting flow.

The specific form of the dependence (2.12) is set in the course of so-called calorimetric measurements, but for the majority of gases this dependence has one and the same distinctive features. Geometrical representation of the dependence (2.12) has the form of a two-dimensional surface in a three-dimensional space of variables (p, υ, T). Figure 2.6 shows isotherms of real gases representing intersections of this surface with planes $T = \text{const}$.

For all gases there exists the so-called critical isotherm, depicted in Figure 2.6 by the heavy line, above and below which the properties of the gas are different. If $T \geq T_{cr}$, where T_{cr} is the critical temperature for a given gas, the gas at any elevation of pressure remains in the gaseous state. If $T < T_{cr}$, then for each temperature T there exists a value of pressure p at which the gas begins to change into the liquid phase, its specific volume being increased from υ' to υ'', after which the resulting medium acquires the properties of a liquid.

The point K is called critical point of a given gas, the quantities $(T_{cr}, p_{cr}, \upsilon_{cr})$ accounting for the individual properties of a gas are constants.

For example, for methane CH_4, which is the major constituent of consists natural gas, $T_{cr} = 190.55$ K and $p_{cr} = 4.641$ MPa. This means that if the gas

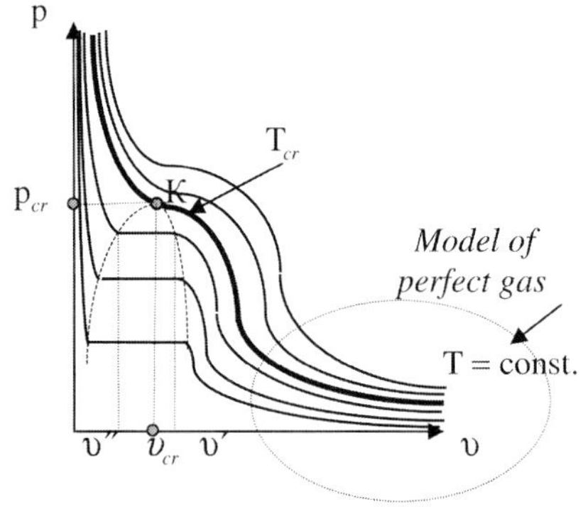

Figure 2.6 Gas isotherms.

temperature exceeds 190.55 K, this gas will be changed into the liquid state without the need for any pressure increase.

2.7.1 Model of a Perfect Gas

If the gas pressure is not too high and the temperature not too low, the isotherms of all gases are similar to each other (see the right-hand part of Figure 2.6 enclosed in the dotted oval line) and with a high degree of accuracy can be approximated by hyperbolas representing the fact that the pressure p is inversely proportional to the specific volume υ.

Under given conditions the interaction of the molecules of a real gas is independent of the form of the molecules, that is of the spatial configuration of the constituent atoms, and is determined only by their total mass. Figuratively speaking, the molecules behave as balls differing in their mass, therefore the number of parameters characterizing the gas decreases from three to one, namely the molecular weight μ_g.

To characterize the thermodynamic state of gases in a given range of pressure and temperature the *model of a perfect gas* is used. Equation (2.11) of the gas state then has the simple form

$$p = \frac{RT}{\upsilon} \quad \text{or} \quad p = \rho RT \tag{2.13}$$

where R is the only constant in the equation and is called the *gas constant* $R = R_0/\mu_g$, where R_0 is the universal gas constant, equal to 8314 J mol^{-1} K^{-1}. Thus all gas constants for a perfect gas depend only on the molecular weight. They are: for methane ($\mu_g \cong 16$ kg kmol^{-1}) $R = 8314/16 \cong 520$ J kg^{-1} K^{-1}); for oxygen O_2 ($\mu_g \cong 32$ kg kmol^{-1}) $R = 8314/32 \cong 260$ J kg^{-1} K^{-1}); for carbon dioxide CO_2 ($\mu_g \cong 44$ kg kmol^{-1}) $R = 8314/44 \cong 189$ J kg^{-1} K^{-1}; for air ($\mu_g \cong 29$ kg kmol^{-1}) $R = 8314/29 \cong 287$ J kg^{-1} K^{-1}.

Equation (2.12) connecting the density, pressure and temperature of a gas is called the *Clapeyron equation*.

The model of a perfect gas operates sufficiently well over a range of not too high pressures and moderate temperatures.

2.7.2 Model of a Real Gas

Despite the fact that the name of this model contains the word "real", one is dealing only with the next, more general, schematization of a gas model. From Figure 2.6 it follows that the hyperbolic dependence (2.13) does not suit observations of gas behavior with increase of pressure and decrease of temperature. Hence, in processes of gas pipeline transportation and underground gas storage, where the pressure may be 5–15 MPa the model of a perfect gas would give improper results were it to be used in calculations.

There is a model of more general form than the model of a perfect gas. This is the *model of a real gas*. One can find the details of such a model for example in Porshakov *et al.*, 2001. Without dwelling on the details of its derivation we note that the mathematical form of this equation is represented as follows

$$p = \frac{Z(p_r, T_r) \cdot RT}{\upsilon} \quad \text{or} \quad p = Z(p_r, T_r) \cdot \rho RT \tag{2.14}$$

differing from Eq. (2.13) by the insertion of the dimensionless factor $Z(p_r, T_r)$ called the *over-compressibility factor*, being a function of two parameters: the *reduced pressure* p_r and the *reduced temperature* T_r, where

$$p_r = \frac{p}{p_{cr}}, \quad T_r = \frac{T}{T_{cr}}.$$

Here p_{cr} and T_{cr} are the critical pressure and temperature.

Hence the model (2.14) takes into account not only the molecular weight of a gas through the constant R but also its thermodynamic parameters such as critical pressure and temperature. It is evident that for moderate values of pressure and temperature $Z \approx 1$ and the model (2.14) transforms into the model of a perfect gas. For a real gas $Z < 1$.

Graphs of the function $Z(p_r, T_r)$ are shown in Figure 2.7.

Exercise. It is required to determine the over-compressibility factor Z of a gas with $p_{cr} = 4.6$ MPa, $T_{cr} = 190$ K at $p = 7.5$ MPa and $T = 288$ K.

Solution. Let us calculate first the reduced parameters of state: $p_{cr} = 1.63$; T_{cr}=1.52. From the plot in Figure 2.7 we determine $Z \cong 0.86$.

There are a lot of approximating formulas to calculate the factor Z. In fact the case is to approximate the equation of state (2.12). However, the properties of a real gas are so complicated that we do not have universal formulas appropriate for all gases over the whole range of governing parameters. Therefore in different cases one should use different approximations. In particular, to simulate processes in gas-pipelines we can use

$$Z(p_r, T_r) \cong 1 - 0.4273 \cdot p_r \cdot T_r^{-3.668} (2.15)$$

or

$$Z(p_r, T_r) \cong 1 - 0.0241 \cdot \frac{p_r}{\theta}, \tag{2.16}$$

where

$$\theta = 1 - 1.68 T_r + 0.78 T_r^2 + 0.0107 T_r^3.$$

It should be particularly emphasized that Eqs. (2.15) and (2.16) are no more than approximations of the state equation of a real gas.

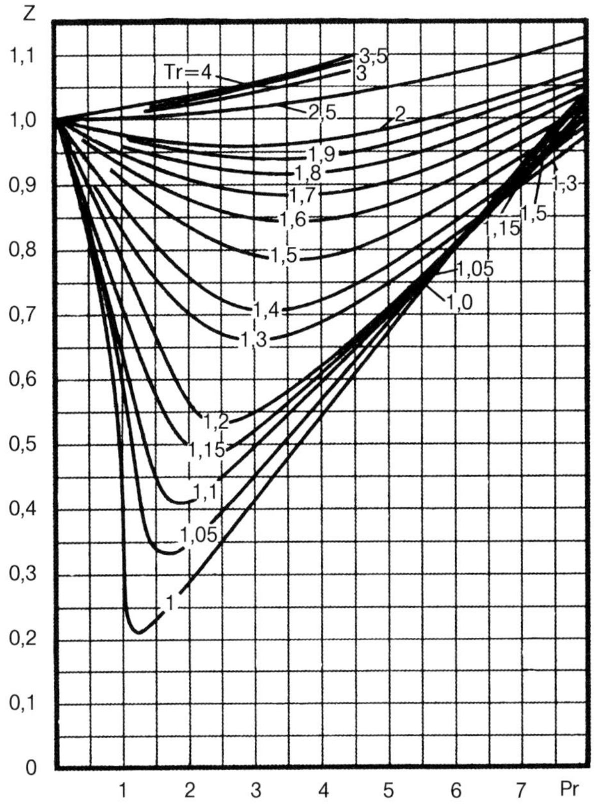

Figure 2.7 Graphs of $Z(p_r, T_r)$ for natural gas.

Exercise. Determine the value of the over-compressibility factor of a gas with $p_{\mathrm{cr}} = 4.6$ MPa, $T_{\mathrm{cr}} = 190$ K at $p = 7.5$ MPa and $T = 288$ K (see the previous example).

Solution. The reduced parameters of state are:

$$p_{\mathrm{r}} = \frac{7.5}{4.6} \cong 1.63; \quad T_{\mathrm{r}} = \frac{288}{190} \cong 1.52.$$

With formulas (2.15) we obtain

$$Z = 1 - 0.4273 \cdot 1.63 \cdot 1.52^{-3.668} \cong 0.850.$$

Similar calculation in accordance with Eq. (2.16) gives:

$$\theta = 1 - 1.68 \cdot 1.52 + 0.78 \cdot 1.52^2 + 0.0107 \cdot 1.52^3 \cong 0.2861,$$

$$Z = 1 - 0.0241 \cdot \frac{1.63}{0.2861} \cong 0.863.$$

It is seen that the error in calculation with Eq. (2.15) is $\approx 2.3\%$, whereas with Eq. (2.16) it is less $\approx 0.8\%$. But this does not mean that the Eq. (2.16) is more precise than Eq. (2.15) because both formulas are only approximations of the state equation of a real gas.

2.8 Model of an Elastic Deformable Pipeline

In the schematization of fluid and gas flow processes in pipelines models of a pipeline are also used.

The simplest model of the pipeline is a *model of non-deformable pipe,* that is of a cylinder with constant invariable internal diameter d_0 and wall thickness δ. The external diameter $D = d_0 + 2\delta$ of the pipeline in this model remains also constant. The model of a non-deformable pipeline appears to be very useful when researching many technological processes of oil and gas transportation.

However, in some cases, for example in studying the phenomenon of *hydraulic hammer,* the model of a non-deformable pipe proves to be inadequate to perceive the point of the phenomenon and it is necessary to use the more complicated model of an *elastic deformable pipeline.*

Experience shows that the volume of the internal space of the pipeline varies insignificantly with changes in temperature and pressure of the transported medium.

In order to account for the volume expansion of a pipeline when the temperature T deviates from its nominal value T_0 one can use

$$V(T) = V_0[1 + 2\alpha_p \cdot (T - T_0)], \tag{2.17}$$

where α_p is the volume expansion factor of the metal from which the pipeline is produced (for steel $\alpha_p \approx 3.3 \cdot 10^{-5}\ \mathrm{K}^{-1}$).

Exercise. How does the volume of the internal space of a steel pipeline section with $D = 530$ mm, $\delta = 8$ mm, $L = 120$ km change during even cooling by 5 K?

Solution. Using Eq. (2.17) we get:

$$V(T) - V_0 = \left(3.14 \cdot \frac{0.514^2}{4}\right) \cdot 2 \cdot (-5) \cdot 3.3 \cdot 10^{-5} \cdot 120\,000 \cdot \cong -8.22,$$

that is the volume decreases by more than by 8 m^3.

The volume of the pipeline internal space is changed to a greater extent with variation in the difference between the internal and external pressures. The simplest formula to calculate changes originated by this phenomenon has been suggested by Joukowsky in his work "*On hydraulic hammer in water-supply pipes*" (1899). Its derivation is illustrated in Figure 2.8.

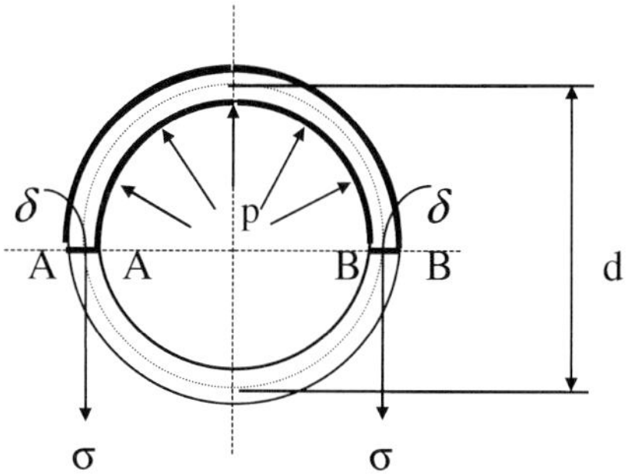

Figure 2.8 Derivation of the formula for the cross-section area change in an elastic deformable pipeline.

The equation of equilibrium of the upper half of the pipe shell shown in Figure 2.8 by a heavy line, under the action of a pressure difference $(p - p_0)$ and circumferential stresses σ resulting in the pipe metal has the following form

$$(p - p_0) \cdot d = \sigma \cdot 2\delta. \tag{2.18}$$

Hooke's law of elasticity as applied to the deformed middle filament shown in Figure 2.8 by a dotted line gives the following relation

$$\sigma = E \cdot \frac{\pi \cdot (d - d_0)}{\pi \cdot d_0} \tag{2.19}$$

where E is the Young's modulus of the pipe material (for steel $E \approx 2 \cdot 10^5$ MPa).

Insertion of σ from Eq. (2.18) into Eq. (2.19) and further replacing factor d by d_0 due to the smallness of the wall thickness as compared to the pipe diameter gives the dependence of the pipe diameter increment $\Delta d = d - d_0$ on the difference $\Delta p = p - p_0$ of the internal and external pressures

$$\Delta d = \frac{d_0^2}{2E\delta} \cdot \Delta p. \tag{2.20}$$

Here d_0 may be taken as the internal diameter of the pipe.

Remark. When deriving Eq. (2.20) it was assumed that axial stresses in the pipe are absent. Such a state of the pipe is called a *plane-stress-state*. But in many cases this assumption is not valid, in particular in steel welded pipelines used in the oil industry there is a *plane-deformable state* in which radial expansion of the pipe generates axial stresses. In such cases Eq. (2.20) should be replaced by the more general formula

$$\Delta d = \frac{d_0^2(1 - \nu_P^2)}{2E\delta} \cdot \Delta p \tag{2.21}$$

where ν_P is the Poisson ratio. However, the correction is insignificant since for steel pipelines $\nu_P^2 \approx 0.078$.

From Eq. (2.20) ensue two useful formulas: one for the increment ΔS of the area of the pipe cross-section and the other for the increment ΔV of the volume of a pipeline section with length L

$$\Delta S = \frac{\pi d_0^3}{4E\delta} \cdot \Delta p, \quad \Delta V = \frac{\pi d_0^3\, L}{4E\delta} \cdot \Delta p \tag{2.22}$$

and associated formulas

$$\Delta S = \frac{\pi d_0^3 \left(1 - \nu_P^2\right)}{4E\delta} \cdot \Delta p, \quad \Delta V = \frac{\pi d_0^3\, L \cdot \left(1 - \nu_P^2\right)}{4E\delta} \cdot \Delta p. \tag{2.23}$$

Exercise. It is required to calculate the increase in the diameter and volume of a section of a steel pipeline with $D = 530$ mm, $\delta = 8$ mm, $L = 120$ km after build up of pressure by 5.0 MPa.

Solution. Using Eq. (2.16) we get

$$\Delta d = \frac{0.514^2}{2 \cdot 2 \cdot 10^{11} \cdot 0.008} \cdot 5 \cdot 10^6 \cong 0.0004 \text{ m} = 0.4 \text{ mm}.$$

From Eq. (2.17) follows

$$\Delta V = \frac{3.14 \cdot 0.514^3 \cdot 120000}{4 \cdot 2 \cdot 10^{11} \cdot 0.008} \cdot 5 \cdot 10^6 \cong 40 \text{ m}^3.$$

For simultaneous deviation of pressure and temperature from their nominal values p_0 and T_0 it is allowable to use the formula

$$V(T) = V_0 \left[1 + 2\alpha_\text{p} \cdot (T - T_0) + \frac{d_0}{E\delta} \cdot (p - p_0)\right]. \tag{2.24}$$

There are also more complicated models of pipelines taking into account the viscous-plastic properties of the pipe material. Such models are applicable for pipelines made from synthetic materials, e.g. from plastics.

3
Structure of Laminar and Turbulent Flows in a Circular Pipe

In the first chapter the flow of a medium transported in a pipeline was considered in the framework of a *one-dimensional model,* that is a model in which the flow is described by characteristics of the fluid velocity, density, pressure, temperature and others averaged over the pipe cross-section. All characteristics of the flow depended only on the longitudinal coordinate x of the cross-section and time t. In such a description additional closure relations reflecting the relations between the average parameters of the flow were needed. For example, relation (1.26) connects the mechanical energy dissipation with the average velocity of the flow, relation (1.29) expresses the tangential friction stress at the internal surface of a pipeline through average parameters of the flow and so on.

It was noted above that to get a closure relation it is necessary to scrutinize not only one-dimensional but also spatial flows taking place in the flow of transported medium inside the pipeline. Let us consider in greater detail such flows with regard to the distribution of the parameters over the pipeline cross-section.

3.1
Laminar Flow of a Viscous Fluid in a Circular Pipe

First let us consider the *laminar flow* of fluid in a circular pipe with radius r_0 (Figure 3.1). Such a flow has only one axial velocity component $u(r)$, dependent on the radial coordinate r equal to the distance from a point under consideration to the pipe wall.

Next separate inside the flow region a cylinder of arbitrary radius r ($r \leq r_0$) and write the balance equation of all the forces acting on the cylinder

$$(p_1 - p_2) \cdot \pi r^2 = |\tau(r)| \cdot 2\pi r L$$

where $\tau(r)$ is the tangential friction stress at the lateral surface of the separated cylinder. From here it follows that the absolute value of the tangential stress τ between the fluid layers is proportional to the cylinder radius r

$$|\tau(r)| = \frac{1}{2} \cdot \frac{\Delta p}{L} \cdot r. \tag{3.1}$$

Modeling of Oil Product and Gas Pipeline Transportation. Michael V. Lurie

ISBN: 978-3-527-40833-7

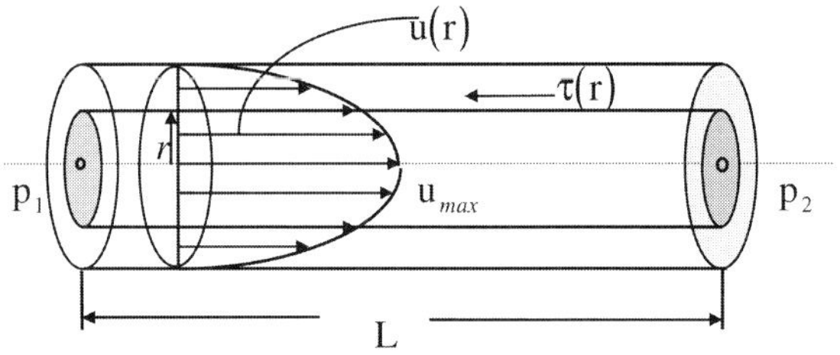

Figure 3.1 Calculation of laminar fluid flow.

If we now replace τ by its expression through the velocity gradient $\mathrm{d}u/\mathrm{d}r$ in accordance with the law of viscous friction (2.2) and take into account that $\tau(r) < 0$, we get the differential equation

$$-\mu\frac{\mathrm{d}u}{\mathrm{d}r} = \frac{1}{2}\cdot\frac{\Delta p}{L}\cdot r \tag{3.2}$$

for the velocity $u(r)$. This equation should be solved for the boundary condition $u = 0$ at $r = r_0$, which is the so-called *sticking condition* (the velocity at the internal surface of the pipe vanishes). As a result we obtain

$$u(r) = u_{\max}\cdot\left(1 - \frac{r^2}{r_0^2}\right); \quad u_{\max} = \frac{\Delta p\cdot r_0^2}{4\,\mu L}. \tag{3.3}$$

It is seen that the fluid velocity distribution over the pipe cross-section has a parabolic form with maximal value $u_{\max}$ at the pipe axis, that is at $r = 0$.

The fluid flow rate Q, that is the fluid volume flowing through a unit cross-section of the pipe in unit time, is equal to

$$Q = \int_0^{r_0} u(r)2\pi r\,\mathrm{d}r = 2\pi u_{\max}\int_0^{r_0} r\cdot\left(1 - \frac{r^2}{r_0^2}\right)\mathrm{d}r = \frac{1}{2}\pi u_{\max} r_0^2$$

or with regard to Eq. (3.3)

$$Q = \frac{r_0^4\pi\Delta p}{8\,\mu L}. \tag{3.4}$$

The relation (3.4) is called the *Poiseuille formula*. It gives the connection between the flow rate of laminar fluid flow in a circular pipe and the pressure drop causing the flow.

Let us introduce the *mean flow-rate-velocity* v which when multiplied by the pipe cross-section area gives the fluid flow rate, that is $v = Q/\pi r_0^2$. Then one obtains very useful formulas

$$v = \frac{r_0^2\Delta p}{8\,\mu L} = 2\cdot u_{\max} \quad \text{or} \quad v = \frac{d^2\Delta p}{32\,\mu L} \tag{3.5}$$

where $d = 2r_0$ is the internal diameter of the pipe.

With the help of these relations it is possible to calculate the tangential stress τ_w at the internal surface of the pipeline. From Eq. (3.1) follows

$$|\tau_w| = \frac{1}{2} \cdot \frac{\Delta p}{L} \cdot r_0.$$

Substitution of $\Delta p/L$ from Eq. (3.5) with regard to Eq. (1.29) yields

$$|\tau_w| = \frac{r_0}{2} \cdot \frac{32\mu v}{d^2} = C_f \cdot \frac{\rho v^2}{2} = \frac{\lambda}{4} \cdot \frac{\rho v^2}{2}$$

From here follows Stokes formula (1.30) for the factor λ of hydraulic resistance in laminar flow of a viscous incompressible fluid in a circular pipe

$$\lambda = \frac{d}{4} \cdot \frac{32\,\mu v}{d^2} \cdot \frac{8}{\rho v^2} = \frac{64}{vd/(\mu/\rho)} = \frac{64}{Re}. \quad (3.6)$$

Exercise. Calculate the hydraulic resistance factor of oil λ ($\nu = 25$ cST) in laminar flow in a circular pipe with diameter 50 mm and flow rate 1 l s^{-1}.

Solution. First determine the mean velocity of the flow

$$v = \frac{Q}{\pi d^2/4} = \frac{0.001}{3.14 \cdot 0.05^2/4} \cong 0.51 \text{ m s}^{-1}.$$

Then calculate the Reynolds number

$$Re = \frac{vd}{\nu} = \frac{0.51 \cdot 0.05}{25 \cdot 10^{-6}} = 1020$$

Since $1020 < Re_{cr} = 2300$, the flow is laminar. In accordance with Eq. (3.6) the hydraulic resistance factor is equal to

$$\lambda = \frac{64}{Re} = \frac{64}{1020} \cong 0.063.$$

3.2 Laminar Flow of a Non-Newtonian Power Fluid in a Circular Pipe

Just as we considered the laminar flow of a Newtonian fluid so we can consider the laminar flow of a non-Newtonian *Ostwald power fluid* in a circular pipe (see Chapter 2). The tangential stress τ for this fluid is related to the shear rate $\dot{\gamma} = du/dr$ by the dependence (2.9)

$$\tau = k \cdot \left|\frac{du}{dr}\right|^{n-1} \cdot \frac{du}{dr}$$

so that the equilibrium Eq. (3.2) takes the form

$$-k \cdot \left|\frac{du}{dr}\right|^{n-1} \cdot \frac{du}{dr} = \frac{1}{2} \cdot \frac{\Delta p}{L} \cdot r$$

Since $\mathrm{d}u/\mathrm{d}r < 0$, the last relation transforms to

$$\frac{\mathrm{d}u}{\mathrm{d}r} = -\left(\frac{\Delta p}{2kL}\right)^{1/n} \cdot r^{\frac{1}{n}}.$$

Integration of this equation with the sticking boundary condition $u(r_0) = 0$ at the pipe wall gives the velocity distribution

$$u(r) = -\frac{n}{n+1} \cdot \left(\frac{\Delta p}{2kL}\right)^{1/n} \cdot \left(r^{\frac{n+1}{n}} - r_0^{\frac{n+1}{n}}\right). \tag{3.7}$$

The maximal velocity of the flow $u_{\max}$ is achieved at the pipe axis $r = 0$ as in the case of a viscous fluid and is equal to

$$u_{\max} = \frac{n}{n+1} \cdot \left(\frac{\Delta p}{2kL}\right)^{1/n} \cdot r_0^{\frac{n+1}{n}}. \tag{3.8}$$

The flow rate of the fluid Q is calculated by the formula

$$\begin{aligned} Q &= 2\pi \int_0^{r_0} r \cdot u(r)\,\mathrm{d}r \\ &= 2\pi \cdot \frac{n}{n+1} \cdot \left(\frac{\Delta p}{2kL}\right)^{1/n} \cdot \int_0^{r_0} r \cdot \left(r_0^{\frac{n+1}{n}} - r^{\frac{n+1}{n}}\right) \mathrm{d}r \end{aligned}$$

Performing integration one gets after some algebra

$$Q = \frac{\pi r_0^3 n}{3n+1} \cdot \left(\frac{r_0 \Delta p/L}{2k}\right)^{\frac{1}{n}}. \tag{3.9}$$

It is reasonable that at $n = 1$, $k = \mu$ Eq. (3.9) converts to Eq. (3.4) known as the Poiseuille law (Wilkenson, 1960).

Introducing the *mean flow-rate-velocity* v as in the previous section and the generalized Reynolds number Re_* according to the relations

$$v = \frac{Q}{\pi r_0^2} \quad \text{and} \quad Re_* = \frac{v^{2-n} \cdot d_0^n}{k/\rho}$$

where $d_0 = 2r_0$, Eq. (3.9) could be written in the habitual form of the Darcy–Weisbach law

$$\frac{\Delta p}{L} = \lambda \cdot \frac{1}{d_0} \cdot \frac{\rho v^2}{2}$$

where the hydraulic resistance factor λ is

$$\lambda = \frac{8 \cdot \left(\frac{6n+2}{n}\right)^n}{Re_*}. \tag{3.10}$$

In the same way Eq. (3.10) at $n = 1$, $k = \mu$ transforms to the Stokes Eq. (1.30) given earlier.

To determine the rheological properties of oil and oil products special devices called viscometers are often used. The most widespread are *capillary viscometers*. Operation of all capillary viscometers is based on the determination of the time of the outflow of a fixed portion of the fluid under test, from the device chamber through a narrow cylindrical tube (capillary). This time is calculated with the use of Eqs. (3.4) and (3.9) replacing in them the pressure gradient $\Delta p/L$ by ρg, where g is the acceleration due to gravity. The flow rate Q in the viscometer of the Ostwald power fluid under consideration takes the following form

$$Q = \frac{\pi {r_0}^3 n}{3n+1} \cdot \left(\frac{r_0 \cdot \rho g}{2k}\right)^{\frac{1}{n}} = \frac{\pi {r_0}^3 n}{3n+1} \cdot \left(\frac{r_0 \cdot g}{2 \cdot k/\rho}\right)^{\frac{1}{n}}. \tag{3.11}$$

Exercise. In order to determine the properties of oil experiments are carried out on a free outflow of a portion (100 ml) of oil from the viscometer chamber. In the first experiment the outflow flows through a cylindrical capillary with internal radius 1 mm and in the second experiment through a similar capillary with internal radius 1.5 mm. In the first case the time of fluid outflow is 1000 s, in the second 180 s. The flow of the oil is modeled by the law of a power fluid. It is required to determine the constants n and k/ρ of the power fluid model.

Solution. From Eq. (3.11) it follows that the ratio of flow rates in both cases is $Q_1/Q_2 = (r_1/r_2)^{3+1/n}$. Since the ratio is inversely proportional to the times of the fluid outflows, we use the equation to determine n, namely $180/1000 = (2/3)^{3+1/n}$. The solution of this equation gives $n \cong 0.81$.

Using further the results of the first experiment we get

$$\frac{0.0001}{1000} = \frac{3.14 \cdot 0.81 \cdot 0.001^3}{3 \cdot 0.81 + 1} \cdot \left(\frac{9.81 \cdot 0.002}{4 \cdot k/\rho}\right)^{1/0.81},$$

from which we get $k/\rho \cong 0.92 \cdot 10^{-6}\ \mathrm{m^2\ s^{-1.19}}$.

3.3 Laminar Flow of a Viscous-Plastic Fluid in a Circular Pipe

Consider the laminar flow of another non-Newtonian fluid – viscous-plastic Bingham fluid (Wilkenson, 1960) (see Chapter 2 and Figure 3.2).

The tangential stress τ in this fluid is related to the shear rate $\dot{\gamma} = du/dr$ by the dependence

$$\tau = \tau_0 + \mu \frac{du}{dr}, \quad (du/dr > 0,\ \tau \geq \tau_0) \tag{3.12}$$

where τ_0 is the limit shear stress. Due to the existence of such stress the flow does not begin immediately after application of the pressure difference to the

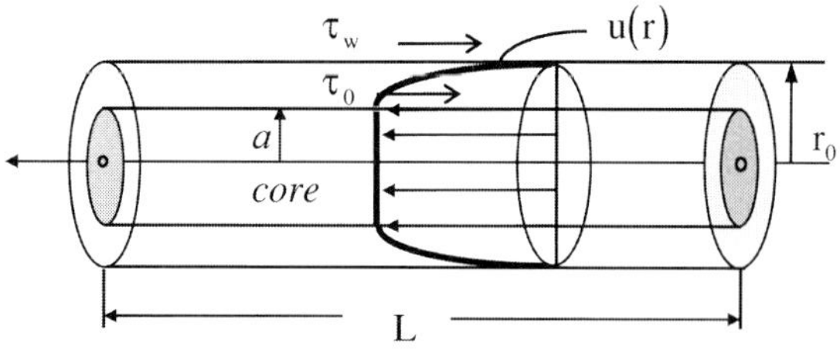

Figure 3.2 A scheme of viscous-plastic fluid flow.

pipe ends but only when this difference exceeds the shear strength, namely when it obeys the following inequality

$$\Delta p \cdot \pi r_0{}^2 > 2\pi r_0\, L \cdot \tau_0 \quad \text{or} \quad \frac{r_0 \cdot \Delta p/L}{2\tau_0} > 1. \tag{3.13}$$

If the condition (3.13) is fulfilled then, near the internal surface of the pipe ($a < r < r_0$), the flow of the fluid starts. On approaching the pipe center where the shear rate decreases the tangential stresses are also reduced and at some distance a from the pipe axis, just where $\tau(a) = \tau_0$, the fluid begins to move as a rigid core. Therefore the region ($0 \le r \le a$) of the flow is called the *flow core*.

It is evident that the following relations are valid

$$\tau(r) = \tau_w \cdot \frac{r}{r_0} \text{ at } a \le r \le r_0,$$

$$\tau(a) = \tau_0 = \tau_w \cdot \frac{a}{r_0} \quad \text{and} \quad \frac{a}{r_0} = \frac{\tau_0}{\tau_w} = \frac{2\tau_0}{r_0 \cdot \Delta p/L}.$$

The velocity distribution in the annular region $a \le r \le r_0$ satisfies the equation

$$\mu \frac{\mathrm{d}u}{\mathrm{d}r} = \tau(r) - \tau_0 = \tau_0 \cdot \left(\frac{r}{a} - 1\right).$$

Integration of this equation over r from $r = r_0$ to an arbitrary r with regard to boundary condition $u(r_0) = 0$ yields

$$u(r) = \frac{r_0 \tau_0}{2\mu}\left[\frac{(r/r_0)^2 - 1}{a/r_0} - 2 \cdot \frac{r}{r_0} + 2\right]. \tag{3.14}$$

In particular the core velocity $u(a)$ is equal to

$$u(a) = -\frac{r_0 \tau_0}{2\mu} \cdot \frac{(1 - a/r_0)^2}{a/r_0}. \tag{3.15}$$

The flow rate Q of the fluid can be obtained using Eqs. (3.14) and (3.15)

$$-Q = 2\pi \int_a^{r_0} r \cdot u(r)\,\mathrm{d}r + \pi a^2 \cdot u(a).$$

Simple rearrangements lead to the following expression

$$Q = \frac{\pi r_0{}^4 \Delta p/L}{8\mu}\left[1 - \frac{4}{3}\cdot\left(\frac{2\tau_0}{r_0\cdot\Delta p/L}\right) + \frac{1}{3}\cdot\left(\frac{2\tau_0}{r_0\cdot\Delta p/L}\right)^4\right]. \tag{3.16}$$

As would be expected this relation at $\tau_0 = 0$ becomes the known Poiseuille formula (3.4) for the flow rate of fluid in a circular pipe (Wilkenson, 1960).

If we introduce two dimensionless parameters $Re = \frac{v\cdot 2r_0}{\nu}$, the Reynolds number and $I = \frac{\tau_0\cdot 2r_0}{\mu v}$, the Ilyushin number and take into account the relation $2\tau_0/(r_0\cdot\Delta p/L) = 8I/\lambda Re$, the equality (3.16) could be represented as

$$\lambda = \frac{64}{Re}\cdot\frac{1}{1 - 4/3\cdot(8I/\lambda Re) + 1/3\cdot(8I/\lambda Re)^4}. \tag{3.17}$$

If $\tau_0 = 0$, then $I = 0$ and consequently $\lambda Re = 64$, that is Eq. (3.17) converts to the known Stokes formula (1.30) for laminar flow of a viscous fluid. In the general case the product λRe depends on the Ilyushin number I. To determine this product one should resolve Eq. (3.17) with respect to λRe for each value of the parameter I (Romanova, 1977).

3.4
Transition of Laminar Flow of a Viscous Fluid to Turbulent Flow

With increase in the velocity of viscous fluid flow in a pipe the laminar flow loses hydrodynamic stability and changes into *turbulent flow*. This flow is characterized by fluctuating motions of the fluid, generation and development of eddies and intensive mixing. A lot of theoretical investigations performed by many outstanding mathematicians and physicists have been devoted to the problem of stability of laminar fluid flows and the determination of the conditions and criteria of the transition from laminar to turbulent flow regimes.

The criterion of transition of laminar flow into a turbulent one is the Reynolds number Re, representing a dimensionless parameter formed by the dimension parameters characterizing the flow and the pipeline: $Re = vd/\nu$, so that at $Re < Re_{cr} = 2300$ the flow is laminar whereas at $Re > Re_{cr} = 2300$ it is turbulent.

The latter conclusion is true when the critical velocity v_{cr} of the flow is determined only by flow parameters such as d, ρ and $\mu (\nu = \mu/\rho)$. In fact, the conditions of transition from laminar to turbulent flow are in many respects determined also by such parameters of the pipeline as the roughness of its internal wall surface Δ. If we accept that $v_{cr} = f(d, \mu, \rho, \Delta)$ then, in accordance with the I-theorem the condition of laminar flow transition to turbulent flow takes the form

$$\frac{v_{cr}\cdot d}{\mu/\rho} = \tilde{f}\left(\frac{\Delta}{d}\right) \quad \text{or} \quad Re_{cr} = \tilde{f}(\varepsilon)$$

where ε is the relative roughness of the pipeline internal wall surface. Thus at $Re < Re_{cr}(\varepsilon)$ the fluid flow is laminar, while at $Re > Re_{cr}(\varepsilon)$ the flow is turbulent. The presence of the additional parameter ε besides the Reynolds number means that there is not a single-valued determined boundary of the transition from laminar to turbulent flow, since the critical Reynolds number depends on the degree of preparation of the pipeline and the fluid for the test experiments. It is known, for example, that in this manner it was possible to lengthen the transition of laminar flow to turbulent flow up to a value of the critical Reynolds number equal to 6000–12 000.

The hydraulic resistance factor λ in the transition region does not have stable values, it is only characterized by drastic lowering. Some formulas for λ in this region will be considered further. The factor λ takes a stable value in establishing the developed turbulent flow in the pipe, that is at $Re > 10\,000$.

3.5
Turbulent Fluid Flow in a Circular Pipe

Consider turbulent flow of a viscous fluid developed in a circular pipe with radius r_0 (Figure 3.3). To describe such a flow let us use the average velocity $u(r)$ representing the true time-averaged velocity of fluid particles passing through a considered point. It is assumed that the resulting velocities are parallel to the pipe axis and independent of the radial coordinate r, that is of the distance between the considered point of the cross-section and the pipe axis.

Similarly, *average tangential stresses* $\tau(r)$ between the fluid layers moving with average velocities $u(r)$ are introduced (Figure 3.3). These stresses are also called *Reynolds stresses,* named in honor of the famous English engineer O. Reynolds who contributed greatly to the development of turbulence theory. The average stress $\tau(r)$ represents the ratio of the friction force acting between macro-layers of turbulent flow separated by a surface element to the area of this element. As in laminar flow, turbulent tangential stresses $\tau(r)$ are suggested to be proportional to the gradients du/dr of average flow velocities

$$\tau(r) = \mu_t \cdot \frac{du}{dr}. \tag{3.18}$$

However, different from laminar flow, the factor μ_t in this relation represents not the intrinsic viscosity of the fluid but the so-called *turbulent dynamic viscosity,* dependent on the structure and mixing intensity of the fluid layers rather than on fluid properties.

The turbulent dynamic viscosity μ_t and the *turbulent kinematic viscosity* $\nu_t = \mu_t/\rho$ of the turbulent flow is caused not by molecular friction of the separated fluid layers, as in laminar flow, but by large-scale fluctuations and momentum

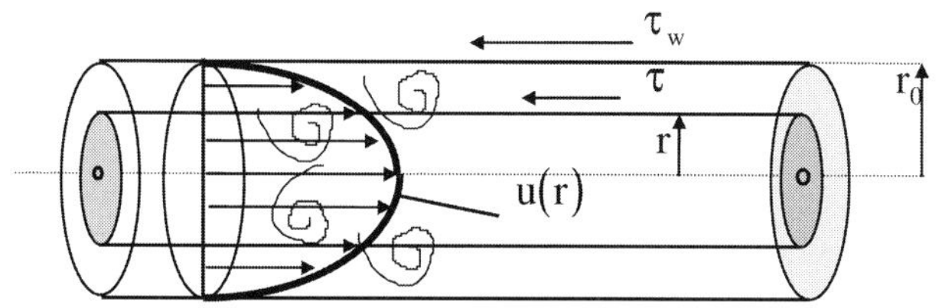

Figure 3.3 The calculation of fluid turbulent flow.

transport with eddies from one macro-layer to another. The transport of momentum is perceived as a friction force acting between these layers. When the layers are moving with equal velocities, that is $du/dr = 0$, then the passing of the momentum from one layer to another is compensated by equal momentum from another layer, therefore $\tau = 0$. If one layer moves faster than another, that is $du/dr > 0$, then there appears between the layers a friction force accelerating the backward layer and retarding the leading one. This means that the leading layer loses more momentum than it obtains from the backward layer. Thus the intensity of momentum exchange between the layers depends on the regime of turbulent flow rather than on the viscosity of the fluid characterized by the factors μ or ν. Hence it follows that the turbulent viscosity μ_t or ν_t is not fluid constant as its molecular analog but depends on parameters of turbulent flow.

Let the law of flow have the following form

$$\frac{1}{\rho}\cdot\tau = \nu_t\cdot\frac{du}{dr} \tag{3.19}$$

It makes sense to suggest that the turbulent viscosity ν_t determined by the structure of turbulent flow be dependent on the parameters of this flow. In accordance with the brilliant idea of Karman it can be accepted that

$$\nu_t = f\left(\nu, \left|\frac{du}{dr}\right|, \left|\frac{d^2u}{dr^2}\right|\right) \tag{3.20}$$

that is, the turbulent viscosity ν_t of the flow depends on the molecular viscosity ν of the fluid itself and parameters absolute values of the first u' and second u'' derivatives of average velocity with respect to the radial coordinate) characterizing the flow.

Since among arguments of the function f there are only two dimensional independent variables ν and u''

$$[\nu] = \frac{L}{T^2}, \quad [|u''|] = \frac{1}{L\cdot T}$$

the dimension of the third argument $|u'|$ could be expressed through the dimensions of the two others as follows (Lurie and Podoba, 1984):

$$[u'] = [\nu]^{1/3}\cdot[|u''|]^{2/3} = \frac{1}{T}.$$

Then owing to the Π-theorem the number of arguments of the function f can be reduced to one and the dependence (3.20) takes the following dimensional form

$$\nu_t/\nu = \tilde{f}\left(\frac{|u'|^3}{\nu \cdot |u''|^2}\right) \Rightarrow \nu_t = \nu \cdot \tilde{f}\left(\frac{|u|^3}{\nu \cdot |u''|^2}\right). \tag{3.21}$$

The formula (3.21) shows that the turbulent viscosity ν_t is equal to the molecular viscosity ν of the fluid multiplied by the dimensionless factor $\tilde{f}$ dependent only on one dimensionless parameter $\eta = |u'|^3/(\nu \cdot |u''|^2)$. Experiments testify, however, that in most of the pipe cross-section, that is in the *flow core*, except for the narrow *wall layer* the turbulent viscosity is practically independent of the molecular viscosity of the fluid. This is because the turbulent viscosity is determined by exchange of momentum between fluid layers due to large-scale eddies rather than by molecular friction. Therefore Eq. (3.21) should have such a structure that the molecular viscosity would have no impact over almost all the cross-section of the flow. The last could be achieved by assuming the function $\tilde{f}$ to be linear, in other words, if $\tilde{f}(\eta) = \kappa \cdot \eta$, where κ is a constant called the *Karman constant*. With regard to the accepted assumption Eq. (3.21) for turbulent viscosity can be rewritten as

$$\nu_t = \nu \cdot \left(\kappa^2 \frac{|u'|^3}{\nu \cdot |u''|^2}\right) = \kappa^2 \cdot \frac{|u'|^3}{(u'')^2}.$$

The law of turbulent friction (3.19) is expressed by

$$\frac{1}{\rho} \cdot \tau = \kappa^2 \frac{|u'|^3}{(u'')^2} \cdot u' \tag{3.22}$$

called the *Karman formula* (Loizyanskiy, 1987). The last equation represents the basic equation of the *phenomenological*, that is resulting from abstract reasoning though adequate for the phenomena under consideration, Karman theory. The constant κ in Eq. (3.22) has been shown by a plethora of experiments to be approximately equal to 0.4 and is a universal constant of the model in the sense that it is the same for all regimes of turbulent flow of fluids in pipes. The relation (3.22) differs significantly from the analogous relation (2.2) for laminar flows.

Let us determine now the distribution $u(r)$ of average velocities of fluid turbulent flow in a circular pipe. First note that tangential stresses $\tau(r)$ as in laminar flow are linearly distributed over the radius

$$\tau(r) = -\frac{1}{2} \cdot \frac{\Delta p}{L} \cdot r$$

since the balance of forces acting on an arbitrarily separated fluid cylinder (see Figures 3.2 and 3.3) is independent of the flow regime (laminar or turbulent). Introduce into the consideration the tangential stress τ_w caused by the friction at the pipe walls

$$\tau_w = \tau(r_0) = -\frac{1}{2} \cdot \frac{\Delta p}{L} \cdot r_0.$$

Then $\tau(r)$ could be expressed through τ_w

$$\tau(r) = \tau_w \cdot \frac{r}{r_0}. \tag{3.23}$$

Substitution instead of $\tau(r)$ from Eq. (3.22) leads to the differential equation

$$\kappa^2 \frac{|u'|^3 u'}{u''^2} = \frac{1}{\rho}\tau_w \cdot \frac{r}{r_0} \tag{3.24}$$

for the velocity distribution $u(r)$ over the radius.

The quantity $(\tau_w/\rho)^{1/2}$ has the dimensions of velocity. It is commonly called the *dynamic velocity* and specified by u_*. In fact it has the sense of a friction stress at the pipe wall since $\rho u_*^2 = |\tau_w|$. In addition the dynamic velocity is related to the above introduced factor λ of hydraulic resistance

$$u_*^2 = \frac{|\tau_w|}{\rho} = \frac{1}{2} C_f \cdot v^2 = \frac{\lambda}{8} \cdot v^2 \Rightarrow \frac{u_*}{v} = \sqrt{\frac{\lambda}{8}}$$

where v is the fluid velocity averaged over the cross-section. Since the factor λ is small ($\lambda \approx 0.01$–0.03), the dynamic velocity $u_* \cong 0.05 \cdot v$, from which follows that the dynamic velocity is 20–25 times smaller than the average flow velocity.

In terms of dynamic viscosity the basic equation (3.24) attains a more compact form

$$\kappa^2 \frac{|u'|^3 u'}{u''^2} = -u_*^2 \cdot \frac{r}{r_0}, \quad (\tau_w < 0). \tag{3.25}$$

Solve this equation for a flow in a circular pipe (Figure 3.3) $u' < 0$ and $u'' < 0$, thus

$$\kappa^2 \frac{|u'^3| u'}{u''^2} = -\kappa^2 \frac{u'^4}{u''^2} = -u_*^2 \cdot \frac{r}{r_0}.$$

From here it ensues that

$$-\frac{u''}{u'^2} = \frac{\kappa}{u_*^2} \cdot \sqrt{\frac{r_0}{r}} \quad \text{or} \quad \frac{d}{dr}\left(\frac{1}{u'}\right) = \frac{\kappa}{u_*^2} \cdot \sqrt{\frac{r_0}{r}}.$$

Repeated integration gives

$$u(r) = \frac{u_*}{\kappa} \cdot \left[\sqrt{\frac{r}{r_0}} + C_1 + C_2 \cdot \ln\left|\sqrt{\frac{r}{r_0}} - C_2\right|\right] \tag{3.26}$$

where C_1 and C_2 are constants of integration.

To determine C_1 and C_2 one needs two *boundary conditions* at the internal pipe wall (Lurie and Podoba, 1984). The first condition is evident. It is the sticking condition in accordance with which the velocity u_w of fluid particles should vanish at the pipe walls

$$u_w = u(r_0) = 0. \tag{3.27}$$

Substituting $r = r_0$ in Eq. (3.26) and taking $u = 0$ we obtain

$$0 = \frac{u_*}{\kappa} \cdot [1 + C_1 + C_2 \cdot \ln|1 - C_2|].$$

Subtracting further the resulting equality term-by-term from Eq. (3.26) we get

$$u(r) = \frac{u_*}{\kappa} \cdot \left[\sqrt{\frac{r}{r_0}} - 1 + C_2 \cdot \ln\left|\frac{\sqrt{r/r_0} - C_2}{1 - C_2}\right|\right]. \tag{3.28}$$

The second condition is more complicated and was not met in the model of laminar flow. The problem is that the differential equation (3.25) for the turbulent model has higher order than the analogous differential equation (3.2) for turbulent flow. Therefore to solve it one needs an additional boundary condition reflecting the interaction of turbulent flow with the pipe walls. Since this condition models the fluid flow in a narrow wall layer it has to connect parameters u'_w and u''_w of the turbulent flow at the pipe wall with the molecular viscosity ν, whose influence is strong in this layer, and the smoothness of the internal surface of the pipe characterized by its absolute roughness Δ. In other words the missing boundary condition should be expressed by the relation

$$G(|u'_w|, |u''_w|, \nu, \Delta) = 0 \tag{3.29}$$

in which the subscript "w" indicates that the corresponding derivatives are calculated at points on the internal pipe surface. The dimension analysis as applied to Eq. (3.29) allows us to rewrite it in dimensionless form

$$\tilde{G}\left(\frac{\nu \cdot u''^2_w}{|u'_w|^3}, \frac{\Delta \cdot |u''_w|}{|u'_w|}\right) = 0.$$

Having resolved the last expression with respect to the first argument, we get

$$\frac{\nu \cdot u''^2_w}{|u'_w|^3} = \tilde{g}\left(\frac{\Delta \cdot |u''_w|}{|u'_w|}\right).$$

For small values of the roughness the right-hand part of this equation could be expanded into a Taylor series leaving in it only the first two terms

$$\frac{\nu \cdot u''^2_w}{|u'_w|^3} = g_0 + g_1 \cdot \frac{\Delta \cdot |u''_w|}{|u'_w|} \tag{3.30}$$

where g_0 and g_1 are dimensionless constants. If we now take into account that at $r = r_0$ in accordance with Eq. (3.25) the relation

$$\kappa^2 \cdot \frac{u_w'^4}{u_w''^2} = u_*^2$$

takes place, meaning that, at the pipe walls, $|u_w''| = \kappa \cdot u_w'^2/u_*$, the condition (3.30) simplifies to

$$\frac{\nu}{|u_w'|^3} \cdot \frac{\kappa^2 \cdot u_w'^4}{u_w''^2} = g_0 + g_1 \cdot \frac{\Delta}{|u_w'|} \cdot \frac{\kappa \cdot u_w'^2}{u_*}$$

and takes the final form ($u_w' < 0$):

$$\nu \cdot u_w' = -\frac{k \cdot u_*^2}{1 + a \cdot (u_* \cdot \Delta/\nu)} \tag{3.31}$$

where $k = g_1/\kappa^2$ and $a = -g_2/\kappa$ are dimensionless constants. These constants are universal like the Karman constant κ, that is, they do not depend on concrete flow and are given once and for all. In other words they are phenomenological constants of the model. A large body of calculation results for different types of flows correlated to each other gives $k = 28$ and $a = 0.31$ (Lurie and Podoba, 1984). Hence, the theory of fluid turbulent flow in a circular pipe is based on three phenomenological constants $\kappa \cong 0.4$; $k \cong 28$ and $a \cong 0.31$.

In the particular case of an ideal smooth internal surface of the pipe ($\Delta \approx 0$), the boundary condition (3.31) reduces to

$$\nu u_w' = -k \cdot u_*^2. \tag{3.32}$$

Thus the boundary condition (3.31) permits us to obtain the second integration constant C_2 in expression (3.28) for the velocity $u(r)$. As a result we have

$$u' = \frac{u_*}{\kappa} \cdot \left[\frac{1}{2\sqrt{r \cdot r_0}} + \frac{C_2}{\sqrt{r/r_0} - C_2} \cdot \frac{1}{2\sqrt{r \cdot r_0}}\right],$$

$$u_w' = u'(r_0) = \frac{u_*}{2r_0 \cdot \kappa} \cdot \frac{1}{1 - C_2}.$$

Insertion of the calculated derivative into condition (3.31) yields the equation

$$-\frac{\nu \cdot u_*}{2r_0 \cdot \kappa} \cdot \frac{1}{1 - C_2} = \frac{k \cdot u_*^2}{1 + a \cdot u_* \Delta/\nu},$$

which gives

$$C_2 = 1 + \frac{1 + a \cdot u_* \cdot \Delta/\nu}{2k\kappa \cdot r_0 u_*/\nu}$$

or, in dimensionless form

$$C_2 = 1 + \frac{1 + a \cdot \varepsilon \cdot Re \cdot u_*/v}{k\kappa \cdot Re \cdot u_*/v} \tag{3.33}$$

where $\varepsilon = \Delta/d$ is the relative roughness; $Re = vd/\nu$ the Reynolds number; v the mean flow velocity; $d = 2r_0$ the pipeline diameter. At the same time note that for turbulent flows with $Re > 10^4$, $\varepsilon < 10^{-3}$, $u_*/v \approx 10^{-2}$ the constant C_2 is close to one,

$$C_2 \approx 1 + \frac{1 + 0.31 \cdot 10^{-3} \cdot 10^4 \cdot 10^{-2}}{28 \cdot 0.4 \cdot 10^4 \cdot 10^{-2}} \cong 1.001. \tag{3.34}$$

Hence, Eq. (3.28) with constant (3.33) gives the distribution of the average velocity of turbulent flow in a circular pipe under the condition that the dynamic velocity u_* is known. The latter in its turn can be expressed through the average flow velocity v and the hydraulic resistance factor λ with Eq. (3.24) as $u_* = v \cdot \sqrt{\lambda/8}$.

Figure 3.4 shows dimensionless turbulent velocity profiles $u(r)/u_{\max}$ related to the maximal value of the fluid velocity at the pipe axis. The lower curve corresponds to $Re = 23\,000$, the top curve to $Re = 3\,200\,000$, the middle one to intermediate values of the Reynolds number (Loitzyanskiy, 1987). The dotted curve depicts the parabola (3.3) giving the velocity distribution in laminar flow regime. The comparison of laminar and turbulent velocity distributions shows that the *turbulent velocity profile has more plane form* and the greater the Reynolds number the more the plane become curves. For the laminar flow regime in accordance with Eq. (3.5) $u_{\max}/v = 2$. For the turbulent regime this ratio is far less. In general it depends on the numbers Re, ε and the average ratio is equal to $u_{\max}/v = 1.15 - 1.25$.

The integration constant C_2 in Eq. (3.33) can be represented as

$$C_2 = 1 + \frac{1 + a \cdot \varepsilon \cdot Re \cdot \sqrt{\lambda/8}}{k\kappa \cdot Re \cdot \sqrt{\lambda/8}}. \tag{3.35}$$

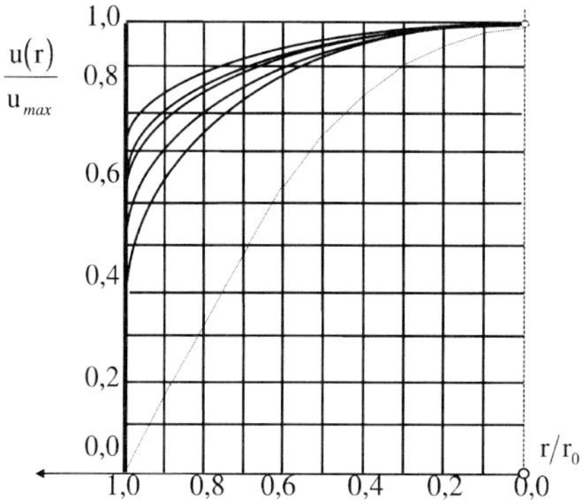

Figure 3.4 Dimensionless average velocity profiles in turbulent flows (Loitzyanskiy, 1987).

On the other hand the average velocity is by definition equal to

$$v = \frac{Q}{\pi \cdot r_0^2} = \frac{1}{\pi \cdot r_0^2} \cdot \int_0^{r_0} 2\pi r \cdot u(r)\,\mathrm{d}r = \frac{2}{r_0^2} \cdot \int_0^{r_0} r \cdot u(r)\,\mathrm{d}r.$$

Substituting here the distribution (3.28) and taking into account the formula (3.35) for C_2 we get the dependence of the hydraulic resistance factor λ on the Reynolds number, that is in fact on the average velocity v, and on the relative roughness ε of the internal surface of the pipe

$$\begin{aligned} v &= \frac{2u_*}{\kappa r_0^2} \int_0^{r_0} r\left[\sqrt{\frac{r}{r_0}} - 1 + C_2 \cdot \ln \frac{C_2 - \sqrt{r/r_0}}{C_2 - 1}\right] dr \\ &= \frac{2u_*}{\kappa} \cdot \int_0^1 \eta \left[\sqrt{\eta} - 1 + C_2 \cdot \ln \frac{C_2 - \sqrt{\eta}}{C_2 - 1}\right] \mathrm{d}\eta \end{aligned} \tag{3.36}$$

where $\eta = r/r_0$. Calculating the integral (3.36) with regard to remark (3.34), we obtain

$$\sqrt{\frac{8}{\lambda}} = \frac{1}{\kappa} \cdot \left(\ln \frac{k\kappa \cdot Re \cdot \sqrt{\lambda/8}}{1 + a \cdot \varepsilon \cdot Re \cdot \sqrt{\lambda/8}} - \frac{137}{60}\right). \tag{3.37}$$

Inserting in Eq. (3.37) the numerical values of the constants k, κ and a we get the equation for the dependence of λ on the numbers Re and ε

$$\frac{1}{\sqrt{\lambda}} = 0.88 \cdot \ln \frac{Re \cdot \sqrt{\lambda}}{1 + 0.11 \cdot \varepsilon \cdot Re \cdot \sqrt{\lambda}} - 0.8 \tag{3.38}$$

called the *universal resistance law*.

1. If we ignore the effect of roughness, that is we assume $0.11 \cdot \varepsilon \cdot Re \cdot \sqrt{\lambda} \ll 1$, e.g. $0.11 \cdot \varepsilon \cdot Re \cdot \sqrt{\lambda} < 0.11$ or $Re \cdot \sqrt{\lambda} < \varepsilon^{-1}$, then the equation for λ becomes

$$\frac{1}{\sqrt{\lambda}} = 0.88 \cdot \ln(Re \cdot \sqrt{\lambda}) - 0.8. \tag{3.39}$$

Its approximate solution takes the form

$$\lambda = \frac{0.3164}{\sqrt[4]{Re}} \tag{3.40}$$

mentioned above and called the *Blasius formula*. The estimation of allowable Reynolds numbers can be conducted as follows

$$Re \cdot \sqrt{\lambda} \cong 0.56 \cdot Re^{7/8} < \frac{1}{\varepsilon} \Rightarrow Re < 1.93 \cdot \varepsilon^{-8/7} \approx 2 \cdot \varepsilon^{-1.143},$$

therefore Eq> (3.39) is true when $Re < 2 \cdot \varepsilon^{-1.143}$.

2. At very large Reynolds numbers ($Re > 10^5$) the equation for λ acquires the form

$$\frac{1}{\sqrt{\lambda}} = 0.88 \cdot \ln \frac{1}{0.11 \cdot \varepsilon} - 0.8$$

where λ is independent of Re. We have

$$\frac{1}{\sqrt{\lambda}} = \left(0.88 \cdot \ln\frac{1}{0.11} + 0.88 \cdot \ln\frac{1}{\varepsilon} - 0.8\right)^{-2} \Rightarrow$$

$$\lambda \cong \left(0.88 \cdot \ln\frac{1}{\varepsilon} + 1.14\right)^{-2}. \tag{3.41}$$

Exercise 1. It is required to calculate the hydraulic resistance factor λ in the flow of diesel fuel with $\nu = 4$ cSt in a pipeline with $d = 500$ mm, $\Delta = 0.25$ mm. The flow rate of the fluid is $Q = 1000\ \mathrm{m^3\ h^{-1}}$.

Solution. Determine the average velocity v of the flow

$$v = \frac{4Q}{\pi d^2} = \frac{4 \cdot 1000/3600}{3.14 \cdot 0.5^2} = 1.415\ \mathrm{m\ s^{-1}}$$

the Reynolds number

$$Re = \frac{vd}{\nu} = \frac{1.415 \cdot 0.5}{4 \cdot 10^{-6}} = 176\,875$$

and the relative roughness

$$\varepsilon = \frac{\Delta}{d} = \frac{0.25}{500} = 0.0005$$

As a result we get the transcendental equation for λ

$$\frac{1}{\sqrt{\lambda}} = 0.88 \cdot \ln\frac{176\,875 \cdot \sqrt{\lambda}}{1 + 0.11 \cdot 0.0005 \cdot 176\,875 \cdot \sqrt{\lambda}} - 0.8$$

We look for the solution of this equation by the method of successive approximations. Consider the function

$$\Phi(\lambda) = \frac{1}{\sqrt{\lambda}} - 0.88 \cdot \ln\frac{176\,878\sqrt{\lambda}}{1 + 9.728 \cdot \sqrt{\lambda}} + 0.8$$

representing the difference in both parts of the resulting equation. We have

$$\lambda = 0.02, \qquad \Phi(0.02) = -0.279;$$

$$\lambda = 0.019, \qquad \Phi(0.019) = -0.086;$$

$$\lambda = 0.018, \qquad \Phi(0.018) = 0.123;$$

$$\lambda = 0.0185, \qquad \Phi(0.0185) = 0.016;$$

$$\lambda = 0.0186, \qquad \Phi(0.0186) = -0.004.$$

It is seen that $\lambda \cong 0.0186$.

Exercise 2. It is required to calculate the hydraulic resistance factor λ in the flow of benzene with $\nu = 0.6$ cSt in a pipeline with $d = 361$ mm, $\Delta = 0.2$ mm. The flow rate of the fluid is $Q = 500\ \mathrm{m^3\ h^{-1}}$.

Solution. Determine the average velocity of the flow

$$v = \frac{4Q}{\pi d^2} = \frac{4 \cdot 500/3600}{3.14 \cdot 0.361^2} = 1.358 \text{ m s}^{-1},$$

the Reynolds number

$$Re = \frac{vd}{\nu} = \frac{1.358 \cdot 0.361}{0.6 \cdot 10^{-6}} = 817\,063.$$

and the relative roughness

$$\varepsilon = \frac{\Delta}{d} = \frac{0.2}{361} = 0.00055.$$

As a result we get the transcendental equation for λ

$$\frac{1}{\sqrt{\lambda}} = 0.88 \cdot \ln \frac{817\,063 \cdot \sqrt{\lambda}}{1 + 49.4 \cdot \sqrt{\lambda}} - 0.8.$$

We look for the solution of this equation by the method of successive approximations. Consider the function

$$\Phi(\lambda) = \frac{1}{\sqrt{\lambda}} - 0.88 \cdot \ln \frac{817\,063\sqrt{\lambda}}{1 + 49.4 \cdot \sqrt{\lambda}} + 0.8,$$

representing the difference between both parts of the resulting equation. We have

$$\lambda = 0.02, \quad \Phi(0.02) = -1.36;$$

$$\lambda = 0.018, \quad \Phi(0.018) = -0.17;$$

$$\lambda = 0.016, \quad \Phi(0.016) = 0.29;$$

$$\lambda = 0.017, \quad \Phi(0.017) = 0.049;$$

$$\lambda = 0.0172, \quad \Phi(0.0172) = 0.003.$$

It is seen that $\lambda \cong 0.0172$.

Turbulent Flows of Non-Newtonian Fluids

For the turbulent flow of a power fluid (see Section 2.6) the following expression for the universal resistance law has been obtained by Dodge and Metzner, 1959

$$\frac{1}{\sqrt{\lambda}} = \frac{0.88}{n^{3/4}} \ln \left[Re \cdot \left(\frac{\lambda}{4} \right)^{1-n/2} \right] - \frac{0.4}{n^{1.2}} \tag{3.42}$$

where n is the exponent in the Ostwald rheological law. This equation may be obtained on the basis of the Karman model (3.25), if in the boundary condition (3.32) we take the factor k to be dependent on the exponent n, that is $k = k(n)$. As is known k is equal to 28 for a Newtonian viscous fluid.

Romanova (1989) gave the resistance law for a power fluid in the form

$$\frac{1}{\sqrt{\lambda}} = \frac{0.88}{n}\ln\left[k(n)Re\cdot\left(\frac{\lambda}{8}\right)^{1-n/2}\right] - 2.83,$$

where $k(n) = 2^{1-n/2}\cdot\exp\left[n\cdot(2.83 - 0.2/n^{1.2})/0.88\right]$. In the same publication the following formulas were suggested as approximate solutions of the above equation

$$\lambda Re^{1/3} = 0.698\cdot n - 1.94\cdot 10^{-2} \quad \text{at } 0.2 \le n \le 0.5,$$

$$\lambda Re^{1/4} = 0.353\cdot n - 3.80\cdot 10^{-2} \quad \text{at } 0.5 < n \le 1.25,$$

$$\lambda Re^{1/5} = 0.234\cdot n - 5.13\cdot 10^{-2} \quad \text{at } 1.25 < n \le 2.0.$$

In all these formulas the generalized Reynolds number is defined as $Re = v^{2-n}\cdot d^n/k$, where k is the kinematic consistency.

The universal resistance law for a viscous-plastic fluid was suggested by Potapov (1975)

$$\frac{1}{\sqrt{\lambda}} = \left(1 - \frac{8He}{\lambda\ Re^2}\right)\cdot\left[0.88\ln(Re\sqrt{\lambda}) - 0.8\right] + 2.76\cdot\frac{8He}{\lambda\ Re^2} \qquad (3.43)$$

where $Re = vd/\nu$ is the Reynolds number and $He = \tau_0 d^2/\rho\nu^2$ is the Hedstroem number.

3.6
A Method to Control Hydraulic Resistance by Injection of Anti-Turbulent Additive into the Flow

Friction losses are the key reason for electric energy expenditure on fluid and gas pumping along pipelines. They are caused by the forces of internal friction between the layers of the moving fluid. In laminar and turbulent flows there occurs the so-called *dissipation* of mechanical energy of ordered motion and its transition into the energy of chaotic motion of the fluid particles, in other words into heat. For turbulent flows this transition has a multi-stage character. The mechanical energy of average motion is transformed first into the energy of large-scale eddies of the turbulent medium, then into the energy of the fluctuation motion of small-scale eddies and finally, due to viscosity forces, into the heat energy of the fluid. Therefore engineers and scientists involved in the problem of pipelines have for a long time been interested in methods of governing the turbulent flow structure with the aim of reducing energy losses.

One such method discovered by the English scientist Toms in the late 1940s consists in injection into the turbulent flow of special high-molecular weight additives to lower the hydraulic resistance. This effect is called by the name of the discoverer, the Toms effect.

The mechanism of operation of all varieties of anti-turbulent additives is based on damping turbulent fluctuations near the internal surface of the pipeline by interaction of the long-length molecules of the additive with turbulent eddies generated near the pipeline wall. As a rule this effect is achieved with very small concentrations of additives, measured commonly in ppm (parts per million of the fluid volume to which the additive is added).

Owing to the damping of the *near-wall turbulence* there is a reduction in the flow hydraulic resistance caused by the pipeline wall. Hence, increased pumping efficiency is achieved with conservation of the pressure drop or pressure lowering at pumping stations. The decrease in hydraulic resistance reduces expenditure on electric power by 20–60%.

The best known anti-turbulent additives for oil products are CDR produced by the American company Dupon-Conoco and NECCAD-547 produced by the Finnish company Neste. These products are based on hydrocarbons. The first is equally suitable for benzene and diesel fuel pumping while the second is recommended primarily for diesel fuels. Both anti-turbulent additives were put through production tests on pipelines in Russia.

The use of anti-turbulent additives has some specific restriction: during prolonged action of the additives in turbulent flow they become degraded; their degradation is especially great when passing through pumps. Therefore, when using additives it is necessary to inject a portion of fresh additives into the flow after each pumping station. It is rational to use anti-turbulent additives to build up the carrying capacity of certain pipeline sections, first the limiting ones.

All anti-turbulent additives reduce the hydraulic resistance factor λ. To calculate this factor the *universal resistance law* (3.37) containing the constant k related to the turbulent flow interaction with pipeline walls was used (see boundary conditions (3.31) and (3.32)). It was found that $k \cong 28$.

The effect of the anti-turbulent additive is that *it changes the intensity of the wall turbulence, that is the additive acts on the magnitude of the constant k.* Therefore it is reasonable to accept as a model of turbulent flow with anti-turbulent additive a model with variable constant k (Ishmuchamedov *et al.*, 1999) dependent on the concentration of anti-turbulent additive θ. Thus the universal quantity k which was taken earlier as constant would be, in the presence of anti-turbulent additive, a function of θ, that is $k = k(\theta)$. In the absence of anti-turbulent additive ($\theta = 0$) then $k(0) = 28$.

The results of tests of anti-turbulent additive CDR have given the dependence $k(\theta)$ shown in Table 3.1.

Table 3.1 Dependence $k(\theta)$ for CDR.

θ, ppm	20	30	40	50	60	70	80	90
$k(\theta)$	61.4	95.1	143	187	249	276	340	380

Table 3.2 Dependence $k(\theta)$ for NECCAD-547.

θ, ppm	40	60	100	180
$k(\theta)$	50	75	150	340

The dependence $k(\theta)$ for anti-turbulent additive NECCAD-547 is shown in Table 3.2.

Remark. The data cited in Tables 3.1 and 3.2 could be improved by changing the anti-turbulent additive composition.

Exercise 1. The pumping of diesel fuel with anti-turbulent additive CDR with $\theta = 40$ ppm is conducted at Reynolds number 40 000. It is required to calculate the hydraulic resistance factor λ.

Solution. The formula (3.39) at $Re = 40\,000$ and $k(40) = 143$ taken from Table 3.1 gives for λ the following transcendental equation

$$\frac{1}{\sqrt{\lambda}} = 0.88 \cdot \ln(143 \cdot 40\ 000 \cdot \sqrt{\lambda}) - 3.745.$$

Its solution, found by the method of successive approximations, yields $\lambda = 0.0153$. This value is significantly less than 0.0224 corresponding to the value of λ in the case of anti-turbulent additive absence in the flow of oil product at the same Reynolds number. The effect is about 31.7%.

Exercise 2. The pumping of diesel fuel with anti-turbulent additive NECCAD-547 with $\theta = 180$ ppm is conducted at Reynolds number 40 000. It is required to calculate the hydraulic resistance factor λ.

Solution. The formula (3.39) at $Re = 40\,000$ and $k(180) = 340$ taken from Table 3.2 gives for λ the following transcendental equation

$$\frac{1}{\sqrt{\lambda}} = 0.88 \cdot \ln(340 \cdot 40\ 000 \cdot \sqrt{\lambda}) - 3.745.$$

Its solution, found by method of successive approximations, yields $\lambda = 0.0129$. This value is significantly less than 0.0224 corresponding to the value of λ in the case of anti-turbulent additive absence in the flow of oil product at the same Reynolds number. The effect is about 42.4%.

In order to select the necessary concentration θ of anti-turbulent additive one can proceed as follows (Ishmuchamedov *et al.*, 1999). From Eq. (3.39) for $k(\theta)$ follows the relation

$$k(\theta) = \frac{1}{Re \cdot \sqrt{\lambda}} \cdot e^{\frac{1+3.745 \cdot \sqrt{\lambda}}{0.88 \cdot \sqrt{\lambda}}} \tag{3.44}$$

which permits the determination of $k(\theta)$ for given λ. Then with the help of Tables 3.1 and 3.2 we obtain the concentration θ of the anti-turbulent additive in the oil product. Multiplying the latter by the total volume of the fluid to be pumped we get the required amount of the anti-turbulent additive.

Exercise 3. It is required at a given pressure resource to increase by 30% the carrying capacity of an oil-pipeline with $D = 377$ mm, $\delta = 8$ mm pumping diesel fuel with $\nu_d = 9$ cSt and flow rate 450 $m^3\,h^{-1}$. Determine the amount of anti-turbulent additive CDR needed to do this.

Solution. Let us calculate the initial values of the pumping rate v_0, Reynolds number Re and the factor of hydraulic resistance λ_0:

$$v_0 = 4Q/S = 4 \cdot 450/(3600 \cdot 3.14 \cdot 0.361^2) = 1.221 \text{ m s}^{-1},$$

$$Re_0 = v_0 d/\nu_d = 1.221 \cdot 0.361/(9 \cdot 10^{-6}) = 48\,976,$$

$$\lambda_0 = 0.0213.$$

Since the carrying capacity must be increased by 30%, the new values of the pumping rate v and the Reynolds number Re will be

$$v = 1.3 \cdot v_0 \cong 1.587 \text{ m s}^{-1}, \Rightarrow Re = 1.3 \cdot Re_0 = 63\,669.$$

Due to the invariability of the pressure resource it should be

$$\lambda_0(Re_0, 0) \cdot v_0^2 = \lambda(Re, \theta) \cdot v^2.$$

This relation gives a new value of λ

$$\lambda = \lambda_0 \cdot (v_0/U)^2 = 0.0213 \cdot (1/1.3)^2 = 0.0126.$$

The constant $k(\theta)$ is determined by Eq. (3.44)

$$k(\theta) = \frac{1}{63\,669 \cdot \sqrt{0.0126}} \cdot e^{\frac{1+3.745\sqrt{0.0126}}{0.88 \cdot \sqrt{0.0126}}} \approx 246.$$

From Table 3.1 we find that this value of k corresponds to $\theta = 60$ ppm of the anti-turbulent additive.

Exercise 4. It is required to increase by 25% at a given pressure resource the carrying capacity of an oil-pipeline section with $D = 530$ mm, $\delta = 8$ mm pumping diesel fuel with $\nu_d = 9$ cSt and flow rate 950 $m^3\,h^{-1}$. Determine the amount of anti-turbulent additive NECCAD-547 needed to do this.

Answer. About 340 ppm.

3.7 Gravity Fluid Flow in a Pipe

The above considered flows of fluid are pertinent to the class of the so-called *enforced (pumped) flows*, since the motion of the flow was forced, that is to overcome the friction force a pressure gradient was needed. However, there are flows in which the primary driving force could be a component of the gravity force. Such flows are called *gravity flows*. A variety of gravity flows are *gravity stratified (divided into layers) flows*. In these flows the fluid moves without completely filling the cross-section, the fluid flows over the lower part of the pipe whereas the upper part of the pipe is filled with vapor and gases evolved from this fluid.

Considering the stratified gravity flow of a viscous incompressible fluid along a section of the pipeline with diameter d and roughness Δ of the internal wall surface inclined at an angle α to the horizontal (Figure 3.5). Our interest is the dependence of the flow rate $Q = v \cdot S$ on the governing parameters, that is the form of the function $Q = f(S, \rho, \mu, g, \sin\alpha, d, \Delta)$. We can write the sought dependence in dimensionless form guided by dimensional theory. Among six arguments there are three dimensionally independent ones, for example d, ρ, μ. Hence the number of independent arguments in dimensional V would be reduced from six to three (Lurie, 2001) and the function $Q = f(S, \rho, \mu, g, \sin\alpha, d, \Delta)$ takes the form

$$\frac{Q/S}{\sqrt{g\sin\alpha \cdot d}} = \tilde{f}\left(\frac{S}{d^2}, \frac{g \cdot \sin\alpha}{\nu^2/d^3}, \varepsilon\right) \tag{3.45}$$

where $\nu = \mu/\rho$ is the kinematic viscosity and $\varepsilon = \Delta/d$ is the relative roughness of the internal surface.

In the theory of gravity fluid flows a parameter R_h, called the *hydraulic radius* (Leibensone *et al.*, 1934), is often introduced. The hydraulic radius is defined as the ratio between the area S of a part of pipe cross-section filled with fluid and the *wetted perimeter* P_s (Figure 3.6).

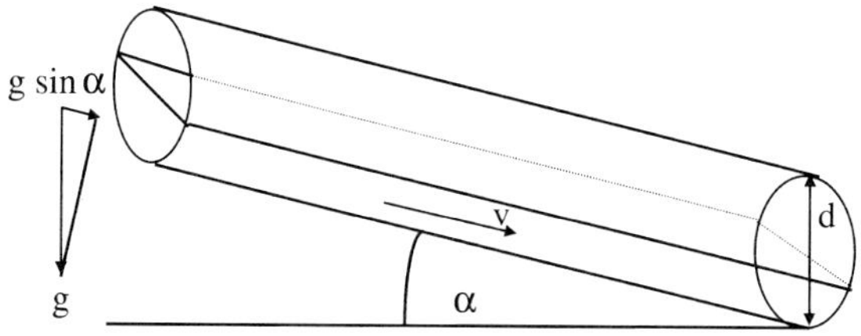

Figure 3.5 A scheme of gravity fluid flow in a pipe.

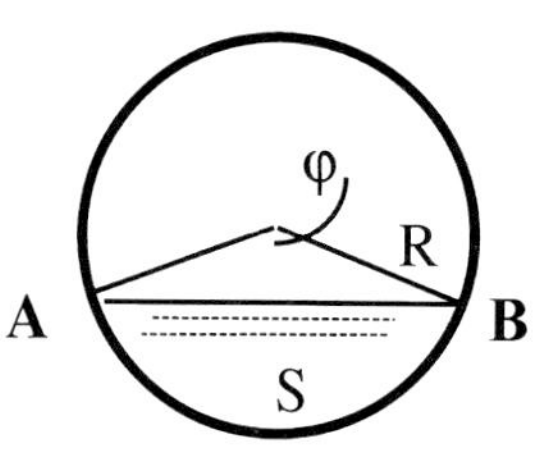

Figure 3.6 Definition of the hydraulic radius.

$$S = \frac{1}{2} \cdot {r_0}^2 \phi - \frac{1}{2} \cdot {r_0}^2 \sin\phi = \frac{1}{2} \cdot {r_0}^2 (\phi - \sin\phi);$$

$$\frac{S}{{r_0}^2} = \frac{1}{2}(\phi - \sin\phi).$$

Since

$$\breve{A}B = r_0 \cdot \phi \quad \text{and} \quad P_s = r_0 \cdot \phi$$

then

$$R_h = \frac{S}{P_S} = \frac{1/2 \cdot {r_0}^2 (\phi - \sin\phi)}{r_0 \cdot \phi} = \frac{r_0}{2} \cdot \left(1 - \frac{\sin\phi}{\phi}\right) \tag{3.46}$$

where φ is the central angle where is seen the part of cross-section filled with fluid and $r_0 = d/2$ the pipe radius. If the pipe is completely filled with fluid ($\varphi = 2\pi$), then $R_h = r_0/2 = d/4$.

Since S/d^2 and R_h/d depend only on the degree of the section filling with fluid, that is on the angle φ, the dependence (3.45) without disturbing generality could be written in equivalent form

$$\frac{Q/S}{\sqrt{g \sin\alpha \cdot R_h}} = \tilde{f}_1 \left(\frac{S}{d^2}, \frac{g \cdot \sin\alpha}{\nu^2/d^3}, \varepsilon\right)$$

or

$$Q = S \cdot C_{Sh} \cdot \sqrt{R_h \cdot \sin\alpha} \tag{3.47}$$

where $C_{Sh} = \sqrt{g} \cdot \tilde{f}_1(S/d^2, \sqrt{gd \sin\alpha} \cdot d/\nu, \varepsilon)$ is the so-called dimensionless *Chezy factor;* $\tilde{f}_1$ is a dimensionless factor dependent on the parameters of the flow regime $\sqrt{gd \sin\alpha} \cdot d/\nu$, degree of fluid filling S/d^2 and the smoothness parameter of the internal surface of the pipe ε (Archangelskiy, 1947 and Christianovitch, 1938).

The formula (3.47) could be represented in a form analogous to the Darcy–Weisbach relation for enforced flow

$$\rho g \sin\alpha = \frac{2 \cdot (d/R_h)}{\tilde{f}_1^2} \cdot \frac{1}{d} \cdot \frac{\rho v^2}{2} \tag{3.48}$$

when in the last expression we replace the pressure gradient $\Delta p/L$ responsible for enforced flow with the rolling down component of the gravity force $\rho g \sin\alpha$ causing motion in the case of gravity flow. Comparison of Eq. (3.48) with the Darcy–Weisbach relation shows that the hydraulic resistance factor λ in stratified gravity flow is related to the factor $\tilde{f}_1$ by the equality

$$\lambda = \frac{2 \cdot (d/R_h)}{\tilde{f}_1^2}$$

from which follows the expression for the Chezy factor

$$C_{\mathrm{Sh}} = \sqrt{\frac{2g \cdot (d/R_{\mathrm{h}})}{\lambda}}. \tag{3.49}$$

There are many empirical formulas for the Chezy factor for pipes with circular as well as non-circular cross-sections (Leibenson *et al.*, 1934). Avoiding detailed treatment of these formulas let us only say that in a first approximation it is possible to use Eq. (3.49) replacing d by $4R_{\mathrm{h}}$, to give $C_{\mathrm{Sh}} = \sqrt{8g/\lambda}$, where $\lambda = \lambda(Re, \varepsilon)$; $Re = 4vR_{\mathrm{h}}/\nu$ and R_{h} the hydraulic radius related to the degree of filling of the pipe cross-section by formulas (3.46).

Pipeline Sections of Gravity Flow

When the pressure in the pipeline section is equal to the saturated vapor tension of the transported fluid, then inside the section continuously appear cavities filled with fluid vapor. In this case the flow could be stratified or could have a more complicated structure in which portions of fluid alternate with vapor-gas cavities (bubbles). The latter flow regime is called *slug flow*.

The flow in a section $[x_1, x_2]$ of the pipeline in which it moves under the action of the gravity force partially (incompletely) filling the pipeline cross-section while the remainder is filled with vapor of this fluid, is called gravity flow. The pressure inside the vapor-gas cavity remains practically invariable and equal to the saturated vapor tension (pressure) p_v. In spite of this the difference in pressures between the sections x_1 (the beginning of the gravity flow section) and x_2 (the end of the gravity flow section) nevertheless exists, it is merely equal to the difference in geometrical heights $(z_1 - z_2)$ of these sections (Figure 3.7).

Stationary gravity flow can exist only on *descending sections of the pipeline*. The beginning of the gravity flow section x_1 is called the *transfer section*. The *transfer section* always coincides with the region of the pipeline profile peak.

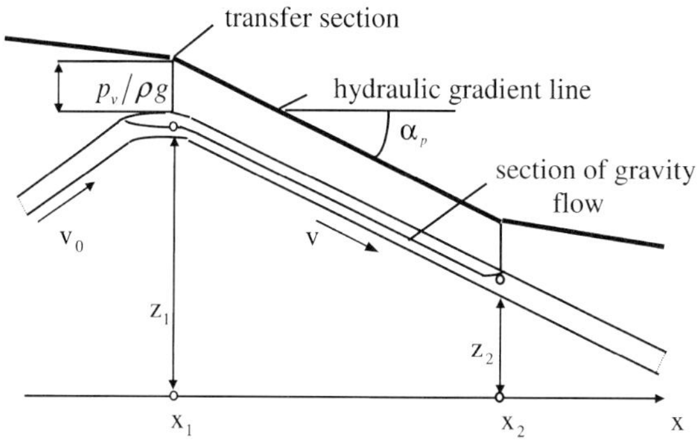

Figure 3.7 A scheme of pipeline gravity flow.

The line of hydraulic gradient of the gravity flow section passes parallel to the pipeline profile at the distance $p_v/\rho g$ over it. Consequently the hydraulic gradient i of the gravity flow section is equal to the slope of the pipeline profile to the horizontal $i = \tan \alpha_p$.

The fluid flow rate in the gravity flow section in the stationary flow regime is equal to the flow rate Q of fluid in the filled sections of the pipeline

$$Q = v_0\, S_0 = v \cdot S \tag{3.50}$$

from which may be concluded that the velocity of fluid flow v in the gravity flow section exceeds the velocity v_0 of the fluid in the pipeline sections filled by the fluid, since the area S of the cross-section part of each gravity flow section filled with the fluid is less than the area S_0 of the complete cross-section of the pipeline, that is $v = v_0 \cdot S_0/S > v_0$.

If the flow of fluid in the gravity flow section is *stratified*, the degree $\sigma = S/S_0$ of pipeline filling with the fluid depends on the ratio $\gamma = i_0/\tan|\alpha_p|$ between the hydraulic gradient $i_0 = \lambda_0 \cdot 1/d \cdot v_0^2/\rho g$ of the pipeline sections completely filled with fluid and the absolute value of the gravity flow section slope α_p to the horizontal. This dependence can be obtained from Eqs. (3.47) or (3.48) solving them with respect to S. To calculate the degree σ of pipeline section filling with fluid it the following approximation formulas have been suggested (Ishmuchamedov *et al.*, 1999):

1. $\sigma = 1$ at $\gamma = i_0/\tan|\alpha_p| \geq 1$. In this case the pipeline cross-section is completely filled with fluid;
2. $\sigma = 1 - 2.98 \cdot 10^{-2} \cdot \sqrt{\frac{2}{\lambda_0}} \cdot (1 - \sqrt{\gamma})$; at $32.32 \cdot \lambda_0 \leq \gamma < 1$;
3. $\sigma = 9.39 \cdot 10^{-2} \cdot \sqrt{\frac{2\gamma}{\lambda_0}} + 0.113$; at $4.87 \cdot \lambda_0 \leq \gamma < 32.32 \cdot \lambda_0$; (3.51)
4. $\sigma = 0.1825 \cdot \left(\frac{2\gamma}{\lambda_0}\right)^{0.356}$; at $\gamma < 4.87 \cdot \lambda_0$.

Exercise. The oil flow rate ($\nu = 8.6$ cSt) in the gravity flow section of an oil-pipeline ($D = 720$ mm, $\delta = 10$ mm, $\Delta = 0.2$ mm) is 900 $m^3\,h^{-1}$. The profile of the section is inclined to the horizontal at an angle $\alpha_p = -1°$. It is required to find the degree of pipeline cross-section filling with oil in this section.

Solution. Calculate the pumping rate v_0, the Reynolds number Re, the factor of hydraulic resistance λ_0 and the hydraulic gradient i_0 in the pumping sections of the pipeline:

$$v_0 = 4 \cdot 900/(3600 \cdot 3.14 \cdot 0.700^2) = 0.650\ \text{m s}^{-1};\ \tan 1° = 0.0175;$$

$$Re = 0.65 \cdot 0.7/(8.6 \cdot 10^{-6}) \cong 52907;\ \lambda_0 = 0.0219;$$

$$i_0 = \lambda_0 \cdot 1/d \cdot v_0^2/(2 \cdot g) = 0.0219 \cdot 1/0.7 \cdot 0.65^2/(2 \cdot 9.81) \cong 0.0007.$$

Determine the parameter γ:

$$\gamma = i_0/\tan|\alpha_p| = 0.0007/0.0175 = 0.04.$$

Since $\gamma = 0.04 < 4.87 \cdot \lambda_0 = 0.1067$, in formulas (3.51) using the fourth case:

$$\sigma = 0.1825 \cdot (2\gamma/\lambda_0)^{0.356} = 0.1825 \cdot (2 \cdot 0.04/0.0219)^{0.356} \cong 0.29,$$

that is, the considered section of the pipeline is approximately 29% filled.

How to determine *whether or not there are gravity flow sections in pipeline sections* under consideration. To answer this question we need to build a combined picture of the pipeline profile and the hydraulic gradient line. If the line of the hydraulic gradient is everywhere over the pipeline profile and the amount by which it exceeds it is the quantity $p_v/\rho g$, where p_v is the saturated vapor pressure of the fluid, then gravity flow sections in the pipeline are absent. If the line of hydraulic gradient at any point approaches the pipeline profile closer than $p_v/\rho g$ or even intersects it, then there exists one or several gravity flow sections in the pipeline (Ishmuchamedov *et al.*, 1999).

Now let us consider Figure 3.8 where a section of pipeline $\mathbf{OO_1}$ is depicted.

Let us begin to build the line $BK_2\Pi_2K_1\Pi_1A$ of the pipeline hydraulic gradient from the end O_1 of the pipeline section under consideration. To do this it is sufficient to know the pressure and the hydraulic gradient at the end of the section. The line of the hydraulic gradient in the segment B K_2 lies significantly over the pipeline profile, therefore cross-sections of the pipeline section are completely filled. However, the line of the hydraulic gradient at the point K_2 approaches the pipeline profile up to a distance $p_v/\rho g$, thus the point K_2 represents the end of the first gravity flow section. Hence, one of the gravity flow sections is found. The line of hydraulic gradient $K_2\Pi_2$ at this section is parallel to the pipeline profile.

Continue to build the line of the hydraulic gradient. It leaves the point Π_2 at an angle whose tangent is equal to the hydraulic gradient, that is in fact parallel to the segment BK_2. It turns out that this line at the point K_1 approaches the pipeline profile at the distance $p_v/\rho g$ for the second time. Consequently, the pressure inside the pipeline again becomes equal to the pressure of the saturated vapor and in the pipeline there should exist vapor-gas cavities. The point K_1 represents the end of the second gravity flow section. Its beginning at the point Π_1 is a transfer section. It is called a transfer section because it is

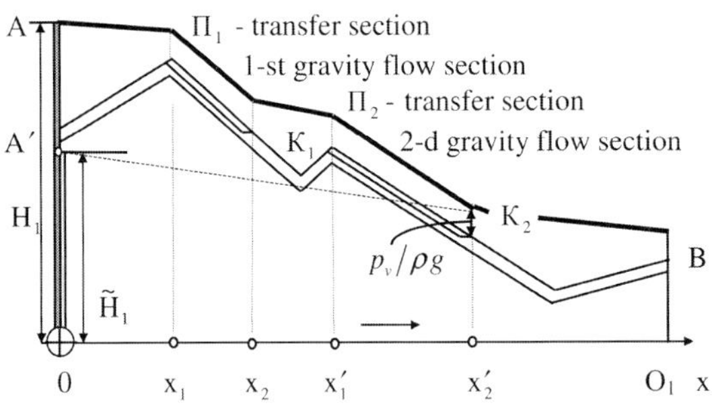

Figure 3.8 A scheme to determine the location of a pipeline gravity flow section.

sufficient to deliver the transported fluid to the point Π_1 so that it reaches then the end O_1 of the section by itself with the help of the gravity flow. Hence, the second gravity flow section $K_1\Pi_1$ is found. The line $K_1\Pi_1$ of the hydraulic gradient at this section passes parallel to the pipeline profile at the distance $p_v/\rho g$ from it.

At the section $\Pi_1 A$ the line of the hydraulic gradient is parallel to its segments BK_2 and $\Pi_2 K_1$ having been built for completely filled pipeline segments.

From Figure 3.8 it follows that the *presence of gravity pipeline sections leads to enhancement of the initial hydraulic head* H_1 *(and consequently the pressure* p_1*) at the station and therefore requires higher expenditures of energy for pumping as compared with a pipeline in which such sections are absent.* If the line of hydraulic gradient beginning from the point K_2 were to be lengthened up to the initial cross-section of the considered pipeline section, it would be possible to determine the hydraulic head $\tilde{H}_1$ needed to pump fluid with the same flow rate in a pipeline of the same length and with the same diameter but without gravity flow sections. It is evident that $H_1 \geq \tilde{H}_1$.

Exercise. Oil ($\rho = 870$ kg m^{-3}, $\nu = 8.5$ cSt, $p_v = 0.02$ MPa) with flow rate 400 m^3 h^{-1} is pumped along an oil-pipeline ($L = 140$ km; $D = 530$ mm, $\delta = 8$ mm, $\Delta = 0.2$ mm). The profile of the section has the form represented in Table 3.3. The pressure at the end of the section is 0.2 MPa. It is required to determine the pressure at the beginning of the section.

Solution. Calculate first the hydraulic gradient.

$$v_0 = 4 \cdot 400/(3600 \cdot 3.14 \cdot 0.514^2) \cong 0.536 \text{ m s}^{-1};$$

$$Re = 0.536 \cdot 0.514/(8.5 \cdot 10^{-6}) = 32\,412;$$

$$\lambda_0 = 0.11 \cdot (0.2/514 + 68/32\,412)^{0.25} \cong 0.025;$$

$$i_0 = 0.025 \cdot 1/0.514 \cdot 0.536^2/(2 \cdot 9.81) \cong 0.00071.$$

Then determine the head losses in the pipeline section between 120 and 140 km. They are $h_{120-140} = i_0 \cdot 20\,000 = 14.2$ m. Therefore the head at the end of the slope, that is at the cross-section $x = 120$ km is equal to $0.2 \cdot 10^6/870 \cdot 9.81 + 14.2 \cong 14.43$ m.

Since $p_v/\rho \cdot g = 20\,000/(870 \cdot 9.81) \cong 2.34$ m, the pipeline at the cross-section $x = 120$ km is still filled with oil. However, the difference in height at the descending section is 100 m (see the profile of the pipeline), thus it is evident that at some cross-section the pressure of oil will be equal to the saturated vapor pressure of oil p_v, so that a part of the descending pipeline

Table 3.3

x, km	0	80	120	140
z, m	100	100	0	0

section will inevitably become a gravity flow section. It is evident that the beginning of this section coincides with the beginning of the descent at $x = 80$ km.

The hydraulic gradient at the plain (completely filled) pipeline segment between the section beginning and the 80th kilometer is equal to the hydraulic gradient at the completely filled pipeline segment between 120 and 140 km, that is 0.71 m km^{-1}, so that the loss of hydraulic pressure is $h_{0-80} \cong 56.8$ m. Therefore the pressure p_1 at the beginning of the section is equal to $870 \cdot 9.81 \cdot 56.8 \cong 484\,771$ Pa or $\approx$4.95 atm.

4
Modeling and Calculation of Stationary Operating Regimes of Oil and Gas Pipelines

In this chapter we consider the calculation of stationary operating regimes of pipelines for transportation of oil, oil products and gas. The equations obtained in the first chapter are used as a basis.

4.1
A System of Basic Equations for Stationary Flow of an Incompressible Fluid in a Pipeline

In stationary flow all parameters of the transported fluid at each cross-section of the pipeline remain constant, that is independent of time. Therefore partial derivatives with respect to time $\partial()/\partial t$ in the equations of Section 1.8 should be taken equal to zero. Consider successively the basic equations describing these flows:

1. Continuity equation (1.6) leads to the equation

$$\frac{\mathrm{d}}{\mathrm{d}x}(\rho v S) = 0$$

which means that the mass flow rate $\dot{M}$ of the transported fluid remains constant

$$\dot{M} = \rho v S = \text{const.}$$

If the fluid is incompressible ($\mathrm{d}\rho/\mathrm{d}t = 0$) and homogeneous ($\rho = \text{const.}$) and the pipeline has invariable diameter ($S = \text{const.}$), the velocity of the fluid would be the same at each cross-section of the pipeline

$$v = \text{const.} \tag{4.1}$$

2. The momentum equation (1.10) gives

$$\rho v \frac{\mathrm{d}v}{\mathrm{d}x} = -\frac{\mathrm{d}p}{\mathrm{d}x} - \frac{4}{d}\tau_w - \rho g \cdot \sin\alpha(x).$$

Modeling of Oil Product and Gas Pipeline Transportation. Michael V. Lurie

ISBN: 978-3-527-40833-7

With regard to the condition $v = \text{const.}$ the momentum equation is simplified and yields

$$\frac{\mathrm{d}p}{\mathrm{d}x} = -\frac{4}{d}\tau_w - \rho g \cdot \sin\alpha(x).$$

If we take in the last equation

$$\tau_w = \frac{\lambda(Re, \varepsilon)}{4} \cdot \frac{\rho v|v|}{2} \text{ and } \sin\alpha(x) = \frac{\mathrm{d}z}{\mathrm{d}x},$$

the momentum equation is transformed into

$$\frac{\mathrm{d}}{\mathrm{d}x}\left(\frac{p}{\rho g} + z\right) = -\lambda(Re, \varepsilon) \cdot \frac{1}{d} \cdot \frac{v|v|}{2g}. \tag{4.2}$$

Note that the Bernoulli equation (1.19) leads also to Eq. (4.2) if we replace in it the hydraulic gradient i_0 by $4|\tau_w|/\rho g d = \lambda(Re, \varepsilon) \cdot v^2/2gd$ in accordance with Eq. (1.23).

3. The equation of total energy balance (1.36) for stationary flow has the form

$$\rho v S \cdot \frac{\mathrm{d}}{\mathrm{d}x}\left(e_{in} + \frac{p}{\rho} + gz\right) = \pi d \cdot q_n$$

or

$$\rho v \frac{\mathrm{d}e_{in}}{\mathrm{d}x} = \frac{4}{d} \cdot q_n - \rho v \cdot \frac{\mathrm{d}}{\mathrm{d}x}\left(\frac{p}{\rho} + gz\right).$$

If in this equation we take $e_{in} = C_v \cdot T + \text{const.}$,

$$\frac{\mathrm{d}}{\mathrm{d}x}\left(\frac{p}{\rho} + gz\right) = -\lambda(Re, \varepsilon) \cdot \frac{1}{d} \cdot \frac{v|v|}{2} \text{ and } q_n = -K \cdot (T - T_{ex})$$

the equation of total energy balance takes the form

$$\rho v C_v \frac{\mathrm{d}T}{\mathrm{d}x} = -\frac{4K}{d}(T - T_{ex}) + \lambda(Re, \varepsilon)\frac{1}{d} \cdot \frac{\rho|v|^3}{2}. \tag{4.3}$$

The system of equations (4.1)–(4.3) with the addition of relations for the hydraulic resistance factor $\lambda(Re, \varepsilon)$ and heat transfer factor K serve as the basis for calculation of the stationary operating regimes of pipelines transporting incompressible fluids, to which class belong oil and oil products.

4.2 Boundary Conditions. Modeling of the Operation of Pumps and Oil-Pumping Stations

The Bernoulli equation in algebraic form results from the continuity equation (4.1) and the differential momentum equation (4.2)

$$\left(\frac{p}{\rho g}+z\right)_{x=0}-\left(\frac{p}{\rho g}+z\right)_{x=L}=\lambda(Re,\varepsilon)\cdot\frac{L}{d}\cdot\frac{v^2}{2g} \tag{4.4}$$

In this relation $x = 0$ and $x = L$ denote, respectively, the initial and terminal cross-sections of the pipeline section with length L. Thus we have one algebraic equation relating three parameters of the flow – the pressure p_0 at the beginning of the pipeline section, the pressure p_L at the end of the pipeline section and the velocity v of the fluid flow. To determine the velocity v or, what is the same, the flow rate of pumping, additional information on the pressures at both ends of the pipeline is needed. This information reflecting the interaction of the considered pipeline section with the rest of the pipeline is introduced into the mathematical model through *boundary conditions*.

In some cases the pressure p_L at the end of the pipeline, that is at $x = L$ could be taken as given, determined e.g. by the conditions required for fluid pumped into reservoirs through a system of intrabase pipelines. Hence, one of the boundary conditions can be a simple condition $p(L) = p_L$.

Another boundary condition models the operation of the oil pumping station (OPS) located at the beginning of the pipeline section, that is at $x = 0$. In fluid flow in the pipeline the pressure gradually decreases because mechanical energy is spent overcoming the force of viscous friction between the fluid layers and is then turned into heat. That is why in a pipeline special *equipment producing pressure* is needed. In general such equipment is called a pump.

4.2.1 Pumps

Pumps represent equipment for compulsory fluid movement from a cross-section with lesser head (line of suction) to a cross-section with greater head (line of discharge). Since the elevations of the pump entrance and exit are, as a rule, identical, one can say that these pumps are *equipment for forced fluid movement from a cross-section with lesser head (line of suction) to a cross-section with greater head (line of discharge)*.

The simplest mathematical model of a pump can be represented as an algebraic equation of the form

$$\Delta H=\frac{p_{\text{ex}}-p_{\text{in}}}{\rho g}=F(Q) \tag{4.5}$$

characterizing the dependence of the differential head ΔH produced by the pump on the fluid flow rate Q. For every actual pump the differential head ΔH appears to be dependent on the fluid flow rate Q called in this case the *feed*. The greater the head produced by a pump, the less, as a rule, is its feed. The dependence $\Delta H = F(Q)$ defines the so-called head-discharge $(Q - H)$ *characteristic of the pump.*

In order to understand the physical basis of this model, let us closely consider the operating principle of one of the commonly encountered pumps, namely the *centrifugal pump*. In centrifugal pumps, used for pumping oil and oil products, the fluid moves from the cross-section with lesser pressure to that with greater pressure under the action of the centrifugal force produced by the rotation of an impeller with profile blades.

Figure 4.1 shows a scheme of a pump impeller with profile blades. Considering the frame of reference related to the rotating impeller, the impeller is believed to be immovable whereas the centrifugal force of inertia $\rho\omega^2 r$, where ρ is the fluid density, ω the angular velocity of the impeller rotation and r the distance of the fluid particle from the rotation axis, acts on the fluid filling the pump.

The centrifugal force causes the fluid to move along the impeller blade from its center to the periphery. This force is capable of overcoming the pressure drop $\Delta p = p_{ex} - p_{in}$, equal to the pressure difference between the pumping pressure p_{ex} at the impeller periphery and the suction pressure p_{in} at the center of the impeller, that is to force the fluid to move from the region of low pressure to the region of high pressure. It is self-evident that to produce such forced motion one needs to spend energy for impeller rotation.

For simplicity let us consider an impeller with radially located blades. The balance equation of forces acting on the fluid moving along the impeller radius from its center to the periphery can be written as:

$$\rho\omega^2 r - \frac{dp}{dr} = \rho \cdot f_\tau(Q)$$

where dp/dr is the radial gradient of pressure opposing the motion and $f_\tau(Q)$ the friction force depending on the discharge Q and increasing with Q.

Integration of the force balance equation over the radius from 0 to R_{im}, where R_{im} is the radius of the impeller, gives

$$\frac{\rho\omega^2 R_{im}{}^2}{2} - \Delta p = R_{im}\rho \cdot f_\tau(Q) \text{ or } \Delta p = \frac{\rho\omega^2 R_{im}{}^2}{2} - R_{im}\rho \cdot f_\tau(Q).$$

Division of both sides of this equation by ρg yields

$$\Delta H = \frac{\omega^2 R_{im}{}^2}{2g} - \frac{R_{im}}{g} \cdot f_\tau(Q). \tag{4.6}$$

Thus the rotation of the impeller with angular velocity ω can force the fluid to move *against the pressure drop* Δp between the periphery and the central

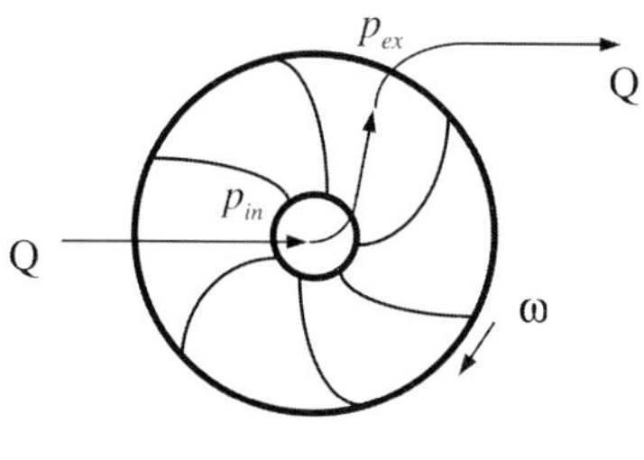

Figure 4.1 Operating principle of a centrifugal pump.

part of the impeller. The maximal value of the pressure drop which the centrifugal force is capable of overcoming is equal to $\rho\omega^2 R_{im}{}^2/2$. This value of Δp is achieved at $Q = 0$ in the absence of a friction force. At $Q > 0$ Eq. (4.6) determining the $(Q - \Delta H)$ characteristic of the pump, where $\Delta H = \Delta p/\rho g$, is obeyed. The pump feed Q decreases with increase in Δp and, conversely, the smaller the pressure drop which the blower has to overcome, the greater the pump feed.

Exercise. It is required to determine the maximal differential head developed by a centrifugal pump with radial located impeller blades having radius 0.25 m and rotating at 3000 rpm.

Solution. 3000 rpm corresponds to the angular velocity $\omega = 2\pi \cdot 3000/60 = 2\pi \cdot 50\ \mathrm{s}^{-1}$. Then, in accordance with Eq. (4.6) we get

$$(\Delta H)_{\max} = \frac{\omega^2 R_b{}^2}{2g} = \frac{4\pi^2 \cdot 50^2 \cdot 0.25^2}{2 \cdot 9.81} \cong 314.4\ \mathrm{m}.$$

The $(Q - \Delta H)$ characteristics of centrifugal pumps operating in stationary regimes are often approximated by the two-term dependence

$$\Delta H = a - b \cdot Q^2 \tag{4.7}$$

where the differential head ΔH is measured in (m) and the flow rate Q in ($\mathrm{m^3\ h^{-1}}$), therefore the dimension of the factor a is (m) and the factor b is ($\mathrm{m/(m^3\ h^{-1})^2}$). For example, the main pump HM 1250-260 produced in Russia rated at a nominal feed of 1250 $\mathrm{m^3\ h^{-1}}$ and nominal head 260 m, has the $(Q - \Delta H)$ characteristic $\Delta H = 331 - 0.451 \cdot 10^{-4} \cdot Q^2$, main pump HM 2500-230 with impeller diameter $D_{im} = 430$ mm rated at nominal feed 2500 $\mathrm{m^3\ h^{-1}}$ and nominal head 230 m has the $(Q - \Delta H)$ characteristic $\Delta H = 280 - 0.792 \cdot 10^{-5} \cdot Q^2$ (ΔH in m, Q in $\mathrm{m^3\ h^{-1}}$) and so on (Vasil'ev *et al.*, 2002).

Figure 4.2 shows the $(Q - H)$ characteristic of the centrifugal pump HM 2500-230.

The two upper curves of this figure represent the $(Q - H)$ characteristics of a pump with accessory impellers (385 and 430 mm, the middle curve shows the dependence of the power consumption N (kW) on the flow rate

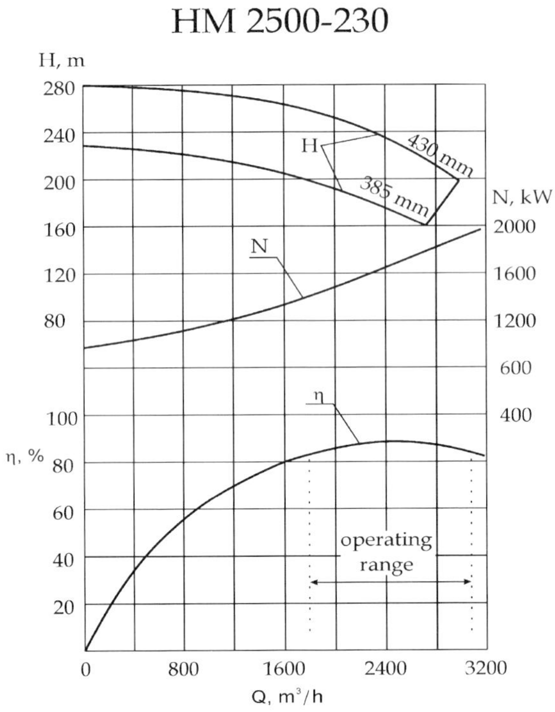

Figure 4.2 Characteristics of the centrifugal pump HM 2500-230.

Q and the bottom curve illustrates the dependence of efficiency $\eta(\%)$ on the flow rate of the transported fluid. In the same figure is also marked the operating range of the pump, that is the range of flow rates Q of the pump. In this range ($1800 < Q < 3000\ \mathrm{m^3\ h^{-1}}$) the efficiency $\eta \approx 85\%$ and the power $N \approx 1600$ kW of the pump have maximal values.

4.2.2 Oil-Pumping Station

Pumps connected *in series or parallel* provide the basis of oil-pumping stations intended to produce driving pressure.

The $(Q - \Delta H)$ characteristics of *pumps connected in series* (Fig. 4.3) are summarized, the fluid flow rates of the pumps are identical $Q_1 = Q_2 = Q$ and the differential heads are given by $\Delta H = \Delta H_1 + \Delta H_2$.

If $\Delta H_1 = a_1 - b_1 \cdot Q^2$ is the characteristic of the first pump and $\Delta H_2 = a_2 - b_2 \cdot Q^2$ the characteristic of the second pump, the characteristic of a system of two pumps connected in series is equal to

$$\Delta H = (a_1 + a_2) - (b_1 + b_2) \cdot Q^2. \tag{4.8}$$

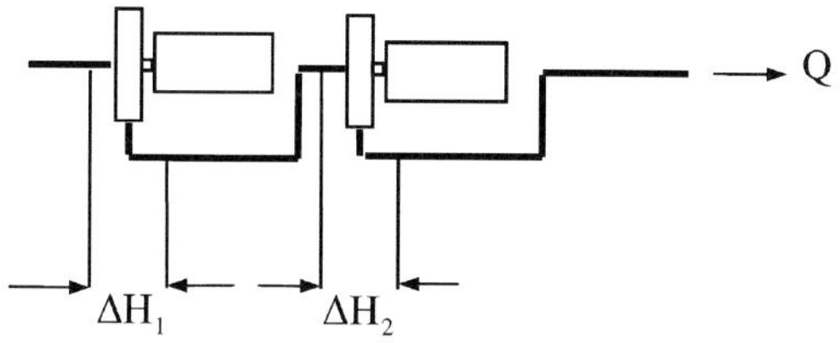

Figure 4.3 Series connection of pumps.

In *parallel connection of pumps* (Fig. 4.4) their $(Q - \Delta H)$ characteristics are different. Fluid discharges in pumps are given by $Q = Q_1 + Q_2$ but the heads produced by each pump are identical $\Delta H = \Delta H_1 = \Delta H_2$.

If $\Delta H = a_1 - b_1 \cdot Q^2$ is the characteristic of the first centrifugal pump and $\Delta H = a_2 - b_2 \cdot Q^2$ that of the second one, the characteristic of the system of two pumps connected in parallel is

$$\sqrt{\frac{(a_1 - \Delta H)}{b_1}} + \sqrt{\frac{(a_2 - \Delta H)}{b_2}} = Q. \tag{4.9}$$

Exercise 1. The $(Q - \Delta H)$ characteristic of a centrifugal pump with impeller diameter 440 mm is $\Delta H = 331 - 0.451 \cdot 10^{-4} \cdot Q^2$. Another pump of the same type but with impeller diameter 465 mm has the $(Q - \Delta H)$ characteristic $\Delta H = 374 - 0.451 \cdot 10^{-4} \cdot Q^2$, ($\Delta H$ in m, Q in $m^3\,h^{-1}$). What characteristic has a system of two pumps connected in series?

Solution. In accordance with Eq. (4.8) we obtain $\Delta H = (331 + 374) - 2 \cdot 0.451 \cdot 10^{-4} \cdot Q^2 = 705 - 0.902 \cdot 10^{-4} \cdot Q^2$.

Exercise 2. The $(Q - \Delta H)$ characteristic of a centrifugal pump with impeller diameter 440 mm is $\Delta H = 331 - 0.451 \cdot 10^{-4} \cdot Q^2$. Another pump of the same type but with impeller diameter 465 mm has the $(Q - \Delta H)$ characteristic $\Delta H = 374 - 0.451 \cdot 10^{-4} \cdot Q^2$, ($\Delta H$ in m, Q in $m^3\,h^{-1}$). It is required to find the characteristic of a system of two pumps connected in parallel?

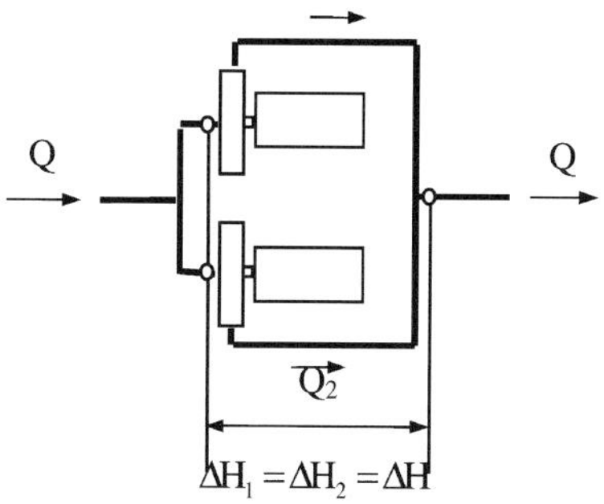

Figure 4.4 Parallel connection of pumps.

Solution. In accordance with Eq. (4.9) we have

$$\sqrt{\frac{(331-\Delta H)}{0.451\cdot 10^{-4}}}+\sqrt{\frac{(374-\Delta H)}{0.451\cdot 10^{-4}}}=Q$$

or

$$\sqrt{331-\Delta H}+\sqrt{374-\Delta H}=6.716\cdot 10^{-3}\cdot Q,\ \text{where}\ \Delta H<331\ \text{m}.$$

The $(Q-\Delta H)$ characteristic of a pumping station is the total $(Q-\Delta H)$ characteristic of all pumps operating in the station (connected in series or parallel) minus the $(Q-\Delta H)$ characteristics of the supply communications. The latter is taken as an element connected in series with the pumps of the station.

Exercise 3. At a pumping station two pumps operate in series with characteristics $\Delta H=F_1(Q)=331-0.451\cdot 10^{-4}\cdot Q^2$ and $\Delta H=F_2(Q)=374-0.385\cdot 10^{-4}\cdot Q^2$. It is also known that the head losses h_c in the station communications, that is the pipeline system of the station, are represented by the dependence

$$h_c=25-0.036\cdot 10^{-4}\cdot Q^2$$

(ΔH, h_c in m, Q in m^3 h^{-1}). It is required to find the characteristic of the pumping station?

Solution. The characteristic of the pumping station $\Delta \mathrm{H}=\mathrm{F(Q)}$ is represented by the sum of the characteristics of the pump system minus head losses in the supply communication $F_1(Q)+F_2(Q)-h_c(Q)$:

$$\Delta H=F(Q)=680-0.800\cdot 10^{-4}\cdot Q^2.$$

Hence, if $(Q-\Delta H)$ the characteristic of the pumping station $\Delta H=F(Q)$ is known, the boundary condition at the initial cross-section $x=0$ of the pipeline section can be the following condition given by this characteristic

$$\frac{p_{ex}-p_{in}}{\rho g}=F(Q)\quad\text{or}\quad\frac{p_0}{\rho g}=\frac{p_u}{\rho g}+\cdot\tilde{F}(v) \tag{4.10}$$

that is, a condition similar to condition (4.5), where $p_0=p_{ex}$ is the pressure at the initial cross-section of the pipeline section, $p_u=p_{in}$ the pressure before the oil-pumping station, called the *head before pumping station*, $F(Q)\equiv F(vS)\equiv\tilde{F}(v)$. When using the two-term dependence of the station differential head ΔH on the flow rate Q the boundary condition (4.10) at the beginning of the pipeline section takes the form

$$\frac{p_0}{\rho g}=\frac{p_u}{\rho g}+a-b\cdot S^2(3600)^2\cdot v^2 \tag{4.11}$$

where the velocity v is measured in (m s^{-1}).

4.3 Combined Operation of Linear Pipeline Section and Pumping Station

To calculate the combined operation of a linear pipeline section and the pumping station located at the beginning of the pipeline section the Bernoulli equation (4.4) is used, in which the pressure $p_0 = p(0)$ at the initial cross-section of the pipeline section is excluded with the help of boundary condition (4.11)

$$\begin{cases} \left(\dfrac{p_0}{\rho g} + z_0\right) - \left(\dfrac{p_L}{\rho g} + z_L\right) = \lambda(Re, \varepsilon) \cdot \dfrac{L}{d} \cdot \dfrac{v^2}{2g}, \\ \dfrac{p_0}{\rho g} = \dfrac{p_u}{\rho g} + (a - 1.296 \cdot 10^7 S^2 b \cdot v^2) \end{cases}$$

where a and b are the approximation factors of the pumping station $(Q - \Delta H)$ characteristic, the velocity v is measured in (m s^{-1}). After eliminating p_0 from these equations we obtain

$$\frac{p_u}{\rho g} - \frac{p_L}{\rho g} + (z_0 - z_L) + a - 1.296 \cdot 10^7 S^2 b \cdot v^2 = \lambda(Re, \varepsilon) \cdot \frac{L}{d} \cdot \frac{v^2}{2g}. \tag{4.12}$$

This equation is called the *head balance equation*. At given values of the head before the pumping station p_u and pressure at the pipeline section end p_L Eq. (4.12) serves to determine the unknown velocity v of the fluid flow in the pipeline.

To solve Eq. (4.12) it is convenient to rearrange all the terms containing the unknown velocity v on the right-hand side of the equation leaving on the left-hand side only the given quantities

$$\frac{p_u}{\rho g} - \frac{p_L}{\rho g} + (z_0 - z_L) + a = \left(\lambda \cdot \frac{L}{d} \cdot \frac{1}{2g} + 1.296 \cdot 10^7 \ S^2 b\right) \cdot v^2.$$

This equation could be solved by the method of successive approximation (*iteration method*). We can demonstrate it with exercises.

Exercise 1. Two identical pumps connected in series and having identical $(Q - H)$ characteristics $\Delta H = 331 - 0.451 \cdot 10^{-4} \cdot Q^2$, ($\Delta H$ in m, Q in m^3 h^{-1}) are pumping diesel fuel ($\rho = 840$ kg m^{-3}, $\nu = 9$ cSt) along the pipeline section ($D = 530 \times 8$ mm, $L = 120$ km, $\Delta = 0.2$ mm, $z_0 = 50$ m, $z_L = 100$ m). It is required to find the flow rate and pressure at the beginning of the section when the pressure p_L at the end of the section is 0.3 MPa, the head before the pumping station h_u is 30 m and it is known that sections of gravity flows are absent in the pipeline.

Solution. Write Eq. (4.11) of the head balance

$$\frac{p_u}{\rho g} - \frac{p_L}{\rho g} + (z_0 - z_L) + 2a = (\lambda \cdot \frac{L}{d} \cdot \frac{1}{2g} + 1.296 \cdot 10^7 S^2 \cdot 2b)v^2.$$

Insertion of the given data yields

$$30 - \frac{0.3 \cdot 10^6}{840 \cdot 9.81} + 50 - 100 + 2 \cdot 331 = \left[\lambda \cdot \frac{120\,000}{0.530 - 2 \cdot 0.008} \cdot \frac{1}{2 \cdot 9.81} + 1.296 \cdot 10^7 \left(\frac{3.14 \cdot 0.514^2}{4}\right)^2 \cdot 2 \cdot 0.451 \cdot 10^{-4}\right] \cdot v^2$$

and

$$605.6 = v^2 \cdot (11\,899.2 \cdot \lambda + 50.3).$$

If as a first approximation it is accepted that $\lambda^{(1)} = 0.02$ then this equation gives $v = 1.449$ m s^{-1}. We need to verify whether or not the factor λ is correctly taken. To do this let us determine the first Reynolds number

$$Re^{(1)} = \frac{1.449 \cdot 0.514}{(9 \cdot 10^{-6})} = 82\,754.$$

Then with formula (1.31) we get

$$\lambda = 0.11 \cdot \left(\frac{0.2}{514} + \frac{68}{82\,754}\right)^{0.25} \cong 0.0205 > \lambda^{(1)} = 0.02.$$

It is seen that the obtained value of the hydraulic resistance factor should be enhanced.

As a second approximation we take $\lambda^{(2)} = 0.0205$. Then the given equation yields $v = 1.434$ m s^{-1}, after which it is necessary to verify whether the factor λ is correctly taken. We have

$$Re^{(2)} = \frac{1.434 \cdot 0.514}{(9 \cdot 10^{-6})} \cong 81\,897;$$

$$\lambda = 0.11 \cdot \left(\frac{0.2}{514} + \frac{68}{81\,897}\right)^{0.25} \cong 0.0206 \approx \lambda^{(2)} = 0.0205.$$

Thus there is good coincidence between the taken and received factor λ. Hence, $v \cong 1.434$ m s^{-1} and

$$Q = \frac{3.14 \cdot 0.514^2}{4} \cdot 1.434 \cong 0.2974 \text{ m}^3\text{s}^{-1} \text{ or}$$

$$Q = 0.2974 \cdot 3600 \cong 1071 \text{ m}^3 \text{ h}^{-1}.$$

The pressure p_0 at the beginning of the pipeline section is determined with the formula $p_0 = \rho g \cdot [h_u + F(Q)]$. As a result we have

$$p_0 = 840 \cdot 9.81 \cdot \left[30 + 2 \cdot \left(331 - 0.451 \cdot 10^{-4} \cdot 1071^2\right)\right] \cong 4.85 \cdot 10^6 \text{ Pa}$$

or 4.85 MPa.

Answer. $1071\ \text{m}^3\ \text{h}^{-1}$; 4.85 MPa.

Exercise 2. The pumping of crude oil ($\rho = 870\ \text{kg m}^{-3}$, $\nu = 25$ cSt) is being conducted by two pumps: HM 2500–230 with characteristic $\Delta H = 251 - 0.812 \cdot 10^{-5} \cdot Q^2$ and HM 3600-230 with $\Delta H = 273 - 0.125 \cdot 10^{-4} \cdot Q^2$ connected in series and rated at feed $1800\ \text{m}^3\ \text{h}^{-1}$. It is known that the $(Q - \Delta H)$ characteristic of the supply communication of the oil-pumping station has the form $\Delta H = 0.15 \cdot 10^{-4} \cdot Q^2$ (here and above ΔH is in m and Q in $\text{m}^3\ \text{h}^{-1}$). It is required to determine the pumping flow rate under the condition that the oil-pipeline section ($D = 820 \times 10$ mm, $L = 150$ km, $z_0 = 80$ m, $z_L = 120$ m, $h_u = 70$ m, $h_L = 40$ m) has an approximately flat character and that sections of gravity flow are absent. Besides it is known that head losses due to local resistances comprise $\approx 2\%$ of the head losses due to friction.

Solution. The equation of the head balance is

$$\left[80 + 70 + \left(251 - 0.812 \cdot 10^{-5} \cdot Q^2\right) + \left(273 - 0.125 \cdot 10^{-4} \cdot Q^2\right)\right.$$
$$\left. -0.15 \cdot 10^{-4} Q^2\right] - [120 + 40] = 1.02 \cdot \lambda \cdot \frac{150\,000}{0.800} \cdot \frac{v^2}{2 \cdot 9.81}.$$

After simplification this equation takes the form

$$514 = v^2 \cdot (9748 \cdot \lambda + 116.6).$$

This equation is solved by the iteration method.

As a first approximation we take $\lambda^{(1)} = 0.02$. Then the equation gives $v = 1.284\ \text{m s}^{-1}$. Now verify whether λ has been chosen correctly.

$$Re = \frac{1.284 \cdot 0.8}{(25 \cdot 10^{-6})} = 41\,088,$$

$$\lambda = \frac{0.3164}{\sqrt[4]{41\,088}} \cong 0.0222 > \lambda^{(1)} = 0.02.$$

As the second approximation we take $\lambda^{(2)} = 0.0222$. Then the equation yields $v = 1.242\ \text{m s}^{-1}$. Now verify whether λ has been chosen correctly

$$Re = \frac{1.242 \cdot 0.8}{(25 \cdot 10^{-6})} = 39\,744;$$

$$\lambda = \frac{0.3164}{\sqrt[4]{39744}} \cong 0.0224 \approx \lambda^{(2)} = 0.0222.$$

Hence $v = 1.242\ \text{m s}^{-1}$ which is equivalent to $Q \cong 2246\ \text{m}^3\ \text{h}^{-1}$.

Answer. $2246\ \text{m}^3\ \text{h}^{-1}$.

4.4 Calculations on the Operation of a Pipeline with Intermediate Oil-Pumping Stations

Consider a pipeline consisting of n successive sections separated by oil-pumping stations (OPS). The transportation of fluid is performed in the so-called *pump-to-pump regime*. When intermediate fluid dumping and pumping are absent we can write the Bernoulli equation for each section

$$\begin{cases} [z_1 + h_{n1} + F_1(Q)] - [z_2 + h_{n2}] = h_{1-2}(Q), \\ [z_2 + h_{n2} + F_2(Q)] - [z_3 + h_{n3}] = h_{2-3}(Q), \\ \dots\dots\dots\dots\dots\dots\dots\dots\dots\dots\dots \\ [z_n + h_{n,n} + F_n(Q)] - [z_L + h_L] = h_{n-(n-1)}(Q), \end{cases} \tag{4.13}$$

where $\Delta H = F_1(Q), \Delta H = F_2(Q), \dots, \Delta H = F_n(Q)$ are the hydraulic $(Q - \Delta H)$ characteristics of oil-pumping stations; $h_{j-(j-1)}(Q)$ the head losses in the sections between the oil-pumping stations dependent on the pumping flow rate Q; $z_1, z_2, \dots, z_n$ the elevations of the oil-pumping stations; $h_{\mathrm{u},1}, h_{\mathrm{u},2}, \dots, h_{\mathrm{u},n}$ the heads before the oil-pumping stations equal to $h_{\mathrm{u,i}} = p_{\mathrm{u,i}}/(\rho g)$; z_L, $h_L = p_L/(\rho g)$ the elevation and piezometric head at the pipeline end $(x = L)$, respectively.

Equations (4.13) represent a system of n algebraic equations (according to the number of sections) with n unknown quantities: flow rate Q and $(n-1)$ heads $h_{\mathrm{u},j}$ before the intermediate oil-pumping stations.

First consequence (equation of head balance):

Term-by-term summation of the equations of the system (4.13) yields the equation called the balance equation of the heads for the whole pipeline

$$(h_{\mathrm{u},1} - h_L) + \sum_{j=1}^{j=n} F_j(Q) = (z_L - z_1) + \sum_{j=1}^{j=n} h_{j-(j-1)}(Q) \tag{4.14}$$

This equation serves to determine the flow rate Q of the fluid (carrying capacity of the pipeline), with all unknown heads $h_{\mathrm{u},j}$ before the intermediate pumping stations being excluded.

It should be taken into account that the flow rate Q found from Eq. (4.14) can be realized in the considered pipeline only when the heads $h_{\mathrm{u},j}$ of all the intermediate stations are greater than the minimum allowed value assuring pump operation without cavitation and the pressure in all cross-sections of the pipeline is less than the permissible value defined by the pipeline strength.

Second consequence (equation for heads before the oil-pumping stations):

Term-by-term summation of only the first s $(s < n)$ equations of the system (4.13) yields the equation for the heads $h_{\mathrm{u},s}$ before the s-th intermediate pumping station

$$h_{\mathrm{u},s} = h_{\mathrm{u},1} + (z_1 - z_L) + \sum_{j=1}^{j=s} \left[F_j(Q) - h_{j-(j-1)}(Q)\right]. \tag{4.15}$$

The flow rate Q in this equation is assumed to be given, since it can be obtained from Eq. (4.14).

In determining the head losses in the pipeline sections it is necessary to account for the possibility of existing transfer points and segments of gravity flow in these sections (see Section 3.7). Let us illustrate the aforesaid with an exercise.

Exercise. An oil-pipeline with length $L = 450$ km consists of three linear sections the data for which are given in the table below. The head $h_{u.1}$ before the leading oil-pumping station is 50 m and the head h_L at the end of the pipeline is 30 m.

At the beginning of each section there is an oil-pumping station with two identical pumps connected in series the characteristics of which are given in the following table

It is required to determine the carrying capacity of the oil-pipeline when pumping oil ($\rho = 900$ kg m^{-3}, $\nu = 30$ cSt) through it and the heads of the intermediate oil pumping stations.

Solution. The balance equations of the heads for the pipeline sections are

$$\left[50 + 50 + 2 \cdot (251 - 0.812 \cdot 10^{-5} \cdot Q^2)\right] - [60 + h_{u,2}] = \lambda \cdot \frac{150\,000}{0.704} \cdot \frac{v^2}{2 \cdot 9.81},$$

$$\left[60 + h_{u,2} + 2 \cdot (285 - 0.640 \cdot 10^{-5} Q^2)\right] - [70 + h_{u,3}] = \lambda \cdot \frac{180\,000}{0.704} \cdot \frac{v^2}{2 \cdot 9.81},$$

$$\left[70 + h_{u3} + 2 \cdot (236 - 0.480 \cdot 10^{-5} Q^2)\right] - [180 + 30] = \lambda \cdot \frac{120\,000}{0.704} \cdot \frac{v^2}{2 \cdot 9.81}.$$

No.	Length, km	D, mm	δ, mm	z_0, m	z_L, m
1.	150	720	8	50	60
2.	180	720	8	60	70
3.	120	720	8	70	180

No.	Type of pump	(Q – ΔH) characteristic	Positive suction head, m
1.	II 2500-230	$\Delta H = 251 - 0.812 \cdot 10^{-5} Q^2$	40
2.	II 3600-230	$\Delta H = 285 - 0.640 \cdot 10^{-5} Q^2$	40
3.	II 5000-210	$\Delta H = 236 - 0.480 \cdot 10^{-5} Q^2$	40

Here we assume that, due to the invariability of the pipeline diameter, the rate of pumping and the hydraulic resistance factors are identical when passing from one section to another; $h_{u,2}$, $h_{u,3}$ are the unknown heads of the intermediate stations that are to be determined.

Term-by-term summation of the above cited equations gives

$$1434 - 3.864 \cdot 10^{-5} \cdot Q^2 \cong 32\,579 \cdot \lambda v^2 \text{ or}$$
$$1434 = v^2 \cdot (32\,579 \cdot \lambda + 75.8).$$

This equation (balance of heads for the whole pipeline) is solved by the iteration method. We take first $\lambda^{(1)} = 0.02$. Then from the above equation we obtain $v^{(1)} = 1.404 \text{ m s}^{-1}$. Next we verify the correctness of the factor λ:

$$Re = \frac{1.404 \cdot 0.704}{(30 \cdot 10^{-6})} \cong 32947,$$

$$\lambda = \frac{0.3164}{\sqrt[4]{32947}} \cong 0.0234 > 0.02.$$

As the second approximation we take $\lambda^{(2)} = 0.0234$. Then the equation yields $v = 1.308 \text{ m s}^{-1}$. We verify again whether the factor λ is correctly chosen:

$$Re = \frac{1.308 \cdot 0.704}{(30 \cdot 10^{-6})} \cong 30\,694;$$

$$\lambda = \frac{0.3164}{\sqrt[4]{30694}} \cong 0.0239 \approx 0.0234.$$

Thus $v = 1.308 \text{ m s}^{-1}$ or $Q \cong 1832 \text{ m}^3 \text{ h}^{-1}$.

From the first balance equation of heads we determine $h_{u,2}$

$$[50 + 50 + 2 \cdot (251 - 0.812 \cdot 10^{-5} \cdot 1832^2)] - [60 + h_{u,2}]$$
$$= 0.0234 \cdot \frac{150\,000}{0.704} \cdot \frac{1.308^2}{2 \cdot 9.81} \text{ and } h_{u,2} \cong 52.7 \text{ m}.$$

The second balance equation gives $h_{u,3}$

$$[60 + 52.7 + 2 \cdot (285 - 0.640 \cdot 10^{-5} \cdot 1832^2)] - [70 + h_{u,3}]$$
$$= 0.0234 \cdot \frac{180000}{0.704} \cdot \frac{1.308^2}{2 \cdot 9.81} \text{ and } h_{u,3} \cong 48.0 \text{ m}.$$

Both the values for the heads of the intermediate oil-pumping stations comply with the requirement of positive suction head and, therefore, the obtained pumping regime is realizable.

4.5
Calculations on Pipeline Stationary Operating Regimes in Fluid Pumping with Heating

To calculate the stationary operating regime of pipelines performing fluid pumping with heating (*high-temperature pumping*) Eqs. (4.1)–(4.3) are used.

$$\begin{cases} \dfrac{d}{dx}(\rho v S) = 0 \\ \dfrac{d}{dx}\left(\dfrac{p}{\rho g} + z\right) = -\lambda(Re, \varepsilon) \cdot \dfrac{1}{d} \cdot \dfrac{v|v|}{2g} \\ \rho v C_v \dfrac{dT}{dx} = -\dfrac{4K}{d}(T - T_H) + \lambda(Re, \varepsilon) \cdot \dfrac{1}{d} \cdot \dfrac{|v|^3}{2g} \end{cases} \quad (4.16)$$

The first equation gives $\rho v S = \text{const.}$, meaning that the mass flow rate is constant in the stationary operating regime. If we take the pipeline diameter to be invariable $d = d_0 = \text{const.}$, that is we neglect heat expansion of the pipe, and the fluid density to be insignificantly variable ($\rho \approx \text{const.}$), that is we also ignore heat expansion of the fluid, from the constancy of mass flow rate the condition of pumping rate constancy $v \approx v_0 = \text{const.}$ follows.

The latter condition $v \approx v_0 = \text{const.}$ allows us to rewrite the system of differential equations (4.16) as

$$\begin{cases} \dfrac{d}{dx}\left(\dfrac{p}{\rho g} + z\right) = -\lambda(Re, \varepsilon) \cdot \dfrac{1}{d_0} \cdot \dfrac{v_0|v_0|}{2g} \\ \rho v_0 C_v \dfrac{dT}{dx} = -\dfrac{4K}{d_0}(T - T_H) + \lambda(Re, \varepsilon) \cdot \dfrac{1}{d_0} \cdot \dfrac{|v_0|^3}{2g} \end{cases} \quad (4.17)$$

As initial conditions, that is conditions at the initial cross-section $x = 0$ of the pipeline, we can take

$$p(0) = p_0; \; T(0) = T_0 \quad (4.18)$$

signifying that the pressure and temperature at the beginning of the pipeline are known. If it is needed to model the oil-pumping station with given $(Q - \Delta H)$ characteristic $\Delta H = a - b \cdot Q^2$ located at the beginning of the pipeline, the initial conditions at $x = 0$ should be taken as

$$\frac{p_0}{\rho g} = \frac{p_u}{\rho g} + a - b \cdot S_0^2 (3600)^2 \cdot {v_0}^2; \quad T(0) = T_0. \quad (4.19)$$

Here, in accordance with Eq. (4.11) a and b are approximated factors the of $(Q - \Delta H)$ characteristics of OPS; p_u the head before the station; $S_0 = \pi d_0^2/4$.

In the general case the system of equations (4.16) should be integrated numerically or solved using one or other simplification suggestions. One consists in ignoring the heat released in the fluid due to the work of internal friction forces as compared to external heat exchange

$$\lambda(Re, \varepsilon)\frac{1}{d_0} \cdot \frac{|v_0|^3}{2g} \le \frac{4K}{d_0} \cdot |T - T_{ex}|.$$

Then the second equation of system (4.16) can be easily integrated and the solution of this equation yields (see Eq. (1.44))

$$T(x) = T_{ex} + (T_0 - T_{ex}) \cdot \exp\left(-\frac{\pi d_0 \cdot K}{C_v \dot{M}_0} x\right) \tag{4.20}$$

where $\dot{M}_0 = \rho v_0 S_0$ is the mass flow rate; T_{ex} the temperature of the external medium, taken as constant. It should be noted that the fluid velocity v_0 is constant but unknown.

The first equation of the system (4.16) can be represented as

$$\left(\frac{p_0}{\rho g} + z_0\right) - \left(\frac{p_L}{\rho g} + z_L\right) = \int_0^L \lambda(Re, \varepsilon)\, dx \cdot \frac{1}{d_0} \cdot \frac{{v_0}^2}{2g}$$

This equation differs from the standard form of the Bernoulli equation in that it takes into account variability of the factor λ with pipeline length. Really, if $T = T(x)$, then $T \neq$ const., the kinematic viscosity of the fluid ν is not constant, it depends on temperature, therefore the Reynolds number $Re = v_0 d_0/\nu(T) = Re(x)$ and the factor of hydraulic resistance λ are functions of x. This circumstance is taken into account in the Bernoulli equation obtained by integration of λ over the pipeline section length. With regard to Eq. (4.19) we have

$$\frac{p_u}{\rho g} - \frac{p_L}{\rho g} + (z_0 - z_L) + a - 1.296 \cdot 10^7 {S_0}^2 b \cdot {v_0}^2 = \int_0^L \lambda(Re, \varepsilon)\, dx \cdot \frac{1}{d_0} \cdot \frac{{v_0}^2}{2g}$$

The system of equations

$$\begin{cases} \dfrac{p_u}{\rho g} - \dfrac{p_L}{\rho g} + (z_0 - z_L) + a - 1.296 \cdot 10^7 {S_0}^2 b \cdot {v_0}^2 = \displaystyle\int_0^L \lambda(Re, \varepsilon)\, dx \cdot \frac{1}{d_0} \cdot \frac{{v_0}^2}{2g}, \\ T(x) = T_{ex} + (T_0 - T_{ex}) \cdot \exp\left(-\dfrac{\pi d_0 \cdot K}{C_v \dot{M}_0} x\right) \end{cases} \tag{4.21}$$

provides a basis to determine the unknown velocity v_0 and consequently the flow rate of fluid pumping with heating. The kinematic viscosity as a function of temperature $\nu(T)$ is assumed to be given, see for example Eq. (2.3).

Analytical Solution

The system of equations (4.21) has an analytical solution when

$$\nu(T) = \nu_{ex} \cdot e^{-\kappa(T - T_{ex})},$$

where ν_{ex} is the kinematic viscosity of the fluid at the temperature T_{ex} of the external media; κ the dependence factor Eq. (2.3);

$$\lambda = \frac{0.3164}{\sqrt[4]{Re}}$$

that is the flow regime of the fluid corresponds to the flow region of the so-called *hydraulic smooth pipes* (Blasius zone). In this case the integral on the right-hand side of the first equation of the system could be calculated in quadratures

$$\int_0^L \lambda\, dx = \frac{0.3164}{\sqrt[4]{v_0 d_0}} \cdot \int_0^L \nu^{1/4}\, dx = \frac{0.3164}{\sqrt[4]{v_0 d_0}} \cdot \int_0^L \nu_{ex}^{1/4} e^{-\frac{\kappa}{4}(T-T_{ex})}\, dx$$
$$= \lambda_{ex} \cdot \int_0^L e^{-\frac{\kappa}{4}(T-T_{ex})}\, dx, \text{ where } \lambda_{ex} = \frac{0.3164}{\sqrt[4]{v_0 d_0/\nu_{ex}}} \tag{4.22}$$

If we convert to the dimensionless coordinate $\xi = x/L$. Then

$$\int_0^L \lambda\, dx = L \cdot \int_0^1 \lambda\, d\xi = L \cdot \int_{T_0}^{T_L} \lambda \frac{d\xi}{dT}\, dT.$$

If now in Eq. (1.44) we neglect the heat release, that is we take $T_{\otimes} = 0$, then Eq. (1.44) gives

$$\rho v_0 C_v \frac{dT}{d\xi} = -\frac{4KL}{d_0} \cdot (T - T_{ex}) \Rightarrow \frac{d\xi}{dT} = -\frac{\rho v_0 C_v d_0}{4KL} \cdot \frac{1}{T - T_{ex}}.$$

Thus

$$\int_0^L \lambda\, dx = L \cdot \int_{T_0}^{T_L} \lambda \frac{d\xi}{dT}\, dT = -\frac{\rho v_0 C_v d_0}{4KL} \cdot L \cdot \int_{T_0}^{T_L} \frac{\lambda}{T - T_{ex}}\, dT.$$

With regard to Eq. (4.22) we have

$$\int_0^L \lambda\, dx = -\frac{\rho C_v v_0 d_0}{4KL} \cdot L \cdot \lambda_{ex} \int_{T_0}^{T_L} \frac{e^{-\frac{\kappa}{4}(T-T_{ex})}}{T - T_{ex}}\, dT.$$

The integral on the right-hand side can be transformed to

$$\int_{T_0}^{T_L} \frac{e^{-\frac{\kappa}{4}(T-T_{ex})}}{T - T_{ex}}\, dT = \int_{-\frac{\kappa}{4}(T_0-T_{ex})}^{-\frac{\kappa}{4}(T_L-T_{ex})} \frac{e^{\eta}}{\eta}\, d\eta.$$

Now gathering together all the results we obtain the following expression for head losses in a non-isothermal fluid flow

$$\int_0^L \lambda\, dx \cdot \frac{1}{d} \cdot \frac{v_0^2}{2g} = \lambda_{eff} \cdot \frac{L}{d} \cdot \frac{v_0^2}{2g} \tag{4.23}$$

where λ_{eff} is the effective factor of hydraulic resistance determined by

$$\lambda_{eff} = \lambda_{ex} \cdot \frac{1}{m} \cdot [Ei(-k) - Ei(-ke^{-m})]; \quad \lambda_{ex} = \frac{0.3164}{\sqrt[4]{v_0\, d_0/\nu_{ex}}};$$

$$k = \frac{\kappa}{4} \cdot (T_0 - T_{ex}); \; m = \frac{4KL}{\rho C_v v_0 d_0} = \frac{\pi K d_0 \cdot L}{C_v \dot{M}_0}. \qquad (4.24)$$

Here we take into account the equality $T_L - T_{ex} = (T_0 - T_{ex}) \cdot \exp(-m)$ following from the basic formula (1.44) and as $Ei(z)$ it is denoted the *Euler function*

$$Ei(z) = \int_{-\infty}^{z} \frac{e^{\eta}}{\eta}\, d\eta, \qquad (4.25)$$

which is widely encountered in technical applications and for which there are special tables. Some values of this function are listed in Table 4.1.

The basic equation (4.21) to determine the velocity v_0 of non-isothermal fluid flow with regard to Eqs. (4.22)–(4.24) takes the form

$$\frac{p_u}{\rho g} - \frac{p_L}{\rho g} + (z_0 - z_L) + a - 1.296 \cdot 10^7 S_0{}^2 b \cdot v_0{}^2 = \lambda_{eff} \cdot \frac{L}{d_0} \cdot \frac{v_0{}^2}{2g} \qquad (4.24)$$

Exercise. Along a practically horizontal oil-pipeline section ($D = 720 \times 10$ mm, $L = 120$ km) is pumped oil ($\rho = 870$ kg m^{-3}, $C_v = 2000$ J kg^{-1} K^{-1}), $\nu_1 = 5$ cSt at $T_1 = 50\,^\circ$C and $\nu_2 = 40$ cSt at $T_2 = 20\,^\circ$C) with heating. The initial temperature T_0 of the oil is 50 °C, the temperature T_{ex} of the environment is 10 °C. The heat-transfer factor K averaged over the pipeline section is 3.5 W m^{-2} K^{-1}). The pumping is carried out with two pumps connected in series. The characteristic of each pump is $\Delta H = 273 - 0.125 \cdot 10^{-4} Q^2$ (ΔH in m, Q in m^3 h^{-1}, $h_L = h_u$). It is required to find the flow rate of pumping and the temperature of the oil at the section end.

Solution. First determine with Eq. (2.30) the dependence of the oil viscosity on temperature $\nu(T)$. We have

$$\nu(T) = 5 \cdot e^{-\kappa \cdot (T-50)}.$$

Here we take $\nu(50) = 5$ cSt. The second condition $\nu(20) = 40$ cSt yields the equation for the factor κ

$$40 = 5 \cdot e^{-\kappa \cdot (20-50)}$$

from which $\kappa \cong 0.0693$ K^{-1}.

Table 4.1

z	−1.0	−0.8	−0.6	−0.4	−0.2	−0.1	−0.05
$Ei(z)$	−0.22	−0.31	−0.45	−0.70	−1.22	−1.82	−2.47

Writing Eq. (4.24) for the balance of heads

$$2 \cdot [273 - 0.125 \cdot 10^{-4} Q^2] = \lambda_{eff} \cdot \frac{120\,000}{0.7} \cdot \frac{v_0{}^2}{2 \cdot 9.81}.$$

The effective factor λ_{eff} of the hydraulic resistance taking into account its variability with pipeline section length in accordance with Eq. (4.23) is

$$\lambda_{eff} = \frac{0.3164}{\sqrt[4]{v_0 \cdot d_0 / \nu_{ex}}} \cdot \frac{1}{m} \cdot \left[Ei(-k) - Ei(-ke^{-m}) \right]$$

where

$$k = \frac{\kappa}{4} \cdot (T_0 - T_{ex}), \; m = \frac{\pi K d_0 \cdot L}{C_v \dot{M}_0} = \frac{4K \cdot L}{\rho C_v v_0 d_0}.$$

Substituting in the balance equation of heads the expression for the flow rate Q using the velocity v_0 of the fluid flow

$$Q = \frac{3.14 \cdot 0.7^2}{4} \cdot v_0 \cdot 3600$$

and taking into account other conditions we get

$$546 = v_0{}^2 \cdot (8737.4 \cdot \lambda_{eff} + 47.94) \tag{4.26}$$

which can be solved by the iteration method.

First approximation. Let us take $\lambda_{eff}{}^{(1)} = 0.02$. Then from Eq. (4.26) we find the velocity of fluid flow $v_0{}^{(1)} = 1.566 \text{ m s}^{-1}$ and verify the correctness of the obtained value. We have

$$\nu_{ex} = 5 \cdot \exp[-0.0693 \cdot (10 - 50)] = 79.95 \text{ cSt};$$

$$\lambda_{ex} = \frac{0.3164}{\sqrt[4]{1.566 \cdot 0.7/(79.95 \cdot 10^{-6})}} \cong 0.029;$$

$$k = 1/4 \cdot 0.0693 \cdot (50 - 10) = 0.693;$$

$$m = \frac{4 \cdot 3.5 \cdot 120\,000}{1.566 \cdot 0.7 \cdot 870 \cdot 2000} \cong 0.881;$$

$$k \cdot \exp(-m) = 0.693 \cdot \exp(-0.881) \cong 0.287;$$

$$\begin{aligned} \lambda_{eff} &= 0.029 \cdot \frac{1}{0.881} \cdot [Ei(-0.693) - Ei(-0.287)] \\ &= 0.029 \cdot \frac{1}{0.881} \cdot [-0.379 - (-0.939)] \cong 0.0186 < \lambda_{eff} \\ &= 0.02. \end{aligned}$$

Since there is a difference between the taken and calculated values of λ_{eff} we make a second approximation.

Second approximation. Let us take $\lambda_{eff}{}^{(2)} = 0.0186$. From Eq. (4.25) we get the new velocity of fluid flow $v_0{}^{(2)} = 1.611$ m s^{-1} and verify its correctness. We have

$$\kappa = 0.0693\ \text{K}^{-1};\ \nu_{ex} = 79.95\ \text{cSt};$$

$$\lambda_{ex} = \frac{0.3164}{\sqrt[4]{1.611 \cdot 0.7/(79.95 \cdot 10^{-6})}} \cong 0.029;$$

$$k = 1/4 \cdot 0.0693 \cdot (50 - 10) = 0.693;$$

$$m = \frac{4 \cdot 3.5 \cdot 120\,000}{1.611 \cdot 0.7 \cdot 870 \cdot 2000} \cong 0.856;$$

$$k \cdot \exp(-m) = 0.693 \cdot \exp(-0.856) \cong 0.294;$$

$$\begin{aligned}\lambda_{eff} &= 0.029 \cdot \frac{1}{0.856} \cdot [Ei(-0.693) - Ei(-0.294)] \\ &= 0.029 \cdot \frac{1}{0.856} \cdot [-0.379 - (-0.921)] \cong 0.0184 \approx 0.0186 \\ &= \lambda_{eff}{}^{(2)}.\end{aligned}$$

Since the taken and calculated values of the factor λ_{eff} show good coincidence the process of successive approximations ends. Hence, $v_0 \cong 1.611$ m s^{-1} and $Q = 2231$ m^3 h^{-1}. Hence, the temperature of the oil at the pipeline section end in accordance with Eq. (4.21) is $T_L = 10 + (50 - 10) \cdot \exp(-0.856) \cong 27\,^\circ$C.

Answer. 2231 m^3 h^{-1}, 27 °C.

4.6
Modeling of Stationary Operating Regimes of Gas-Pipeline Sections

For modeling the stationary flow of a compressible gas in a gas-pipeline the following equations are used:

- continuity equation (1.6):

$$\frac{d}{dx}(\rho v S) = 0 \Rightarrow \dot{M} = \rho v S = \text{const.} \tag{4.27}$$

Since the gas density ρ decreases with pressure drop, from Eq. (4.27) it follows that the gas velocity v increases from the beginning of the pipeline section to its end;

- momentum equation (1.10):

$$\rho v \frac{dv}{dx} = -\frac{dp}{dx} - \frac{4}{d}\tau_w - \rho g \sin\alpha(x).$$

If the gas velocity v increases, the acceleration $v \cdot dv/dx$ of the gas is distinct from zero. It is evident that the estimation

$$\rho v \frac{dv}{dx} = \frac{d}{dx}(\rho v^2) \ll \frac{d}{dx}(p)$$

is valid when the gas velocity v is small compared to the velocity of sound in a gas c, $c = \sqrt{\gamma RT} \approx \sqrt{1.31 \cdot 500 \cdot 300} \cong 440\ \text{m s}^{-1}$, where $\gamma = C_p/C_v$ is the adiabatic index (for methane $\gamma = 1.31$). So, for example, when the velocity of compressed gas with density $50\ \text{kg m}^{-3}$ varies over Δv the quantity $\rho v \cdot \Delta v$ is $50 \cdot 10 \cdot \Delta v = 500 \cdot \Delta v$ at gas velocity $10\ \text{m s}^{-1}$, whereas the pressure variation Δp calculated with the Joukowski formula $\Delta p = \rho c \Delta v$ is $50 \cdot 440 \cdot \Delta v = 22\,000 \cdot \Delta v$, that is about 45 times greater. Thus, the acceleration of gas and often the gravity component $\rho g \sin\alpha$ in the momentum equation may, as a rule, be neglected. Then the momentum equation of the gas expresses in essence the equality of the driving forces: pressure and friction

$$\frac{dp}{dx} = -\frac{4}{d}\tau_w = -\frac{4}{d} \cdot C_f \frac{\rho v^2}{2} = -\lambda \frac{1}{d} \cdot \frac{\rho v^2}{2}; \tag{4.28}$$

- balance equation of total energy

$$\rho v S \cdot \frac{dJ}{dx} = \pi d \cdot q_n$$

This equation ignores the work due to the force of gravity. If in this equation we use the dependence of enthalpy J on pressure p and temperature T, that is we take $J = J(p, T)$, and the external heat exchange in the form of Newton's law (1.42), then

$$\rho v \left[\left(\frac{\partial J}{\partial T}\right)_p \cdot \frac{dT}{dx} + \left(\frac{\partial J}{\partial p}\right)_T \frac{dp}{dx} \right] = -\frac{4K}{d}(T - T_{ex}).$$

Denoting $(\partial J/\partial T)_p = C_p$ as the specific heat of a gas at constant pressure and $(\partial J/\partial p)_T = -D_* C_p$, where D_* is the Joule-Thompson factor, the last equation takes the form

$$\rho v C_p \frac{dT}{dx} - \rho v C_p D_* \frac{dp}{dx} = -\frac{4K}{d}(T - T_{ex}). \tag{4.29}$$

Using the expression for gas enthalpy J involving internal energy and other parameters of state

$$J = e_{in} + \frac{p}{\rho} = C_v T + Z(p, T) \cdot RT = [C_v + Z(p, T)R] \cdot T.$$

The factors C_p and D_* can be expressed through the factors C_v and $Z(p, T)$ as follows

$$C_p = \left(\frac{\partial J}{\partial T}\right)_p = C_v + R \cdot \left[Z + RT \left(\frac{\partial Z}{\partial T}\right)_p \right];$$

$$D_* = -\frac{1}{C_p}\left(\frac{\partial J}{\partial p}\right)_T = -\frac{RT}{C_p} \cdot \left(\frac{\partial Z}{\partial p}\right)_T. \tag{4.30}$$

From this it is seen that the Joule-Thompson effect (cooling of heat-insulated gas in a gas-pipeline through a pressure drop, $(\partial Z/\partial p)_T < 0$, see Fig. 2.7) manifests itself only for a real gas when $Z \neq 1$ (as a rule the factor $D_* \approx 3\text{–}5$ K MPa^{-1}).

If we neglect the Joule-Thompson effect, the energy balance equation is simplified to

$$\rho v C_p \frac{dT}{dx} = -\frac{4K}{d}(T - T_{ex}) \tag{4.31}$$

and its solution takes the form of Eq. (1.46):

$$\frac{T(x) - T_{ex}}{T_0 - T_{ex}} = \exp\left(-\frac{\pi d \cdot K}{C_p \dot{M}} x\right). \tag{4.32}$$

If we define the average temperature T_{av} of the gas over pipeline section length by the expression

$$T_{av} = \frac{1}{L} \cdot \int_0^L T(x)\, dx$$

and substitute into it the distribution $T(x)$ from Eq. (4.32), we get

$$T_{av} = T_{ex} + \frac{T_0 - T_L}{\ln\left(\dfrac{T_0 - T_{ex}}{T_L - T_{ex}}\right)} \tag{4.33}$$

where the temperature T_L at the end of pipeline section is represented by the expression

$$T_L - T_{ex} = (T_0 - T_{ex}) \cdot \exp\left(-\frac{\pi d \cdot K \cdot L}{C_p \dot{M}}\right)$$

following from Eq. (4.32).

4.6.1 Distribution of Pressure in Stationary Gas Flow in a Gas-Pipeline

This distribution is obtained from Eqs. (4.27) and (4.28)

$$\begin{cases} \rho v S = \dot{M} = const., \\ \dfrac{dp}{dx} = -\lambda \dfrac{1}{d} \cdot \dfrac{\rho v^2}{2} \end{cases} \tag{4.34}$$

and the equation of the gas state $p = Z\rho RT$. Transformation of the second equation (4.32) yields

$$\frac{dp}{dx} = -\lambda \frac{1}{d} \cdot \frac{\rho v^2}{2} = -\frac{1}{2} \cdot \lambda \cdot \frac{\rho^2 v^2 S^2}{d \cdot \rho S^2} = -\frac{1}{2} \cdot \lambda \cdot \frac{\dot{M}^2}{(p/ZRT) \cdot (\pi^2 d^5/16)}.$$

If we take

- $\lambda = \text{const.}$, see e.g. Eq. (1.34);
- $T \approx T_{av} = \text{const.}$;
- $Z(p, T) \approx Z_{av} = \text{const.}$

the resulting differential equation gives after integration the distribution of $p(x)$ along the pipeline section length

$$\frac{p}{2} \cdot \frac{dp}{dx} = -\frac{16}{\pi^2} \cdot \frac{\lambda \cdot Z_{cp} R T_{cp} \cdot \dot{M}^2}{d^5}$$

or

$$p^2(x) = p_0{}^2 - \frac{16\lambda \cdot Z_{av} R T_{av} \dot{M}^2}{\pi^2 d^5} x. \tag{4.35}$$

Here we used the initial condition $p = p_0$ at $x = 0$. Equation (4.33) means that $p^2(x)$ decreases linearly with the pipeline section length

$$p^2(x) = p_0{}^2 - (p_0{}^2 - p_L{}^2) \cdot \frac{x}{L} \tag{4.36}$$

where p_L is the pressure at the section end.

If we define the average pressure p_{av} over the pipeline section length as

$$p_{av} = \frac{1}{L} \cdot \int_0^L p(x)\, dx$$

and we insert it into the distribution of $p(x)$ from Eq. (4.36), we get

$$p_{av} = \frac{2}{3}\left(p_0 + \frac{p_L{}^2}{p_0 + p_L}\right). \tag{4.37}$$

From Eq. (4.36) one gets in particular

- pressure p_L (in Pa) at the end of the pipeline section with length L for given mass flow rate $\dot{M}$

$$p_L{}^2 = p_0{}^2 - \frac{16\lambda \cdot Z_{av} R T_{av} \cdot L}{\pi^2 d^5} \cdot \dot{M}^2 \tag{4.38}$$

or

$$p_L = \sqrt{p_0{}^2 - \frac{16\lambda \cdot Z_{av} R T_{av} \cdot L}{\pi^2 d^5} \cdot \dot{M}^2}; \tag{4.39}$$

- mass flow rate $\dot{M}$ (in kg s^{-1}) for given pressures p_0 and p_L at the beginning and end of the pipeline section, respectively

$$\dot{M} = \frac{\pi}{4}\sqrt{\frac{p_0{}^2 - p_L{}^2}{\lambda \cdot Z_{av} R T_{av} \cdot L} \cdot d^5}. \tag{4.40}$$

If we take $\lambda = 0.067(2\Delta/d)^{0.2}$ in accordance with Eq. (1.34), the mass flow rate is found to be proportional to $d^{2.6}$

$$\dot{M} = A \cdot \sqrt{\frac{p_0{}^2 - p_L{}^2}{L}} \cdot d^{2,6}$$

where A is the proportionality factor.

The mass flow rate $\dot{M}$ of gas (kg s^{-1}) can also be measured by the volume flow rate taken under standard conditions, i.e. at $p_{st} = 101\,325$ Pa; $T_{st} = 293.15$ K. To do this it is sufficient to divide $\dot{M}$ by the density ρ_{st} of gas under standard conditions

$$Q_k = \frac{\dot{M}}{\rho_{st}} = \frac{\rho}{\rho_{st}} \cdot vS = \frac{\rho}{\rho_{st}} \cdot Q_v \tag{4.41}$$

Here ρ and $Q_v = vS$ are the density and volume flow rate of gas at the pipeline cross-section, respectively. The quantity Q_k (m^3 s^{-1}) is the so-called *commercial flow rate* of gas. In fact the commercial flow rate of gas is the mass flow rate expressed in volume units under standard conditions, in other words the mass flow rate in volume calculus. From Eq. (4.41) in particular follows that the commercial flow rate Q_k of gas is ρ/ρ_{st} times greater than the volume flow rate Q_v. The key advantage is that in contrast to the volume flow rate that varies from one cross-section to another, the commercial flow rate remains invariable with the length of the gas-pipeline in stationary flows.

4.6.2
Pressure Distribution in a Gas-Pipeline with Great Difference in Elevations

In this case one has to use the following equations instead of Eq. (4.34)

$$\begin{cases} \rho v S = \dot{M} = \text{const.}, \\ \dfrac{dp}{dx} = -\lambda \dfrac{1}{d} \cdot \dfrac{\rho v^2}{2} - \rho g \cdot \sin\alpha, \quad \alpha = \dfrac{dz}{dx} \end{cases} \tag{4.42}$$

containing the pipeline profile $z(x)$. If we take $T \approx T_{av} = \text{const.}$ and $Z \approx Z_{av} = \text{const.}$, we get

$$\frac{dp(x)^2}{dx} = -\frac{2g}{Z_{av}RT_{av}} \frac{dz}{dx} \cdot p(x)^2 - \frac{\lambda \cdot Z_{av}RT_{av} \cdot \dot{M}^2}{(\pi d^2/4)^2 d} \tag{4.43}$$

to find the function $p(x)$.

The solution of this equation with initial condition $p(0) = p_0$ is

$$p(x)^2 = -\frac{\lambda \cdot Z_{av}RT_{av} \cdot \dot{M}^2}{(\pi d^2/4)^2 d} \cdot \int_0^x \exp\left[\frac{2g}{Z_{av}RT_{av}}(z(\varsigma) - z(x))\right] d\varsigma + p_0{}^2 \times \exp\left[\frac{2g}{Z_{av}RT_{av}}(z_0 - z(x))\right]. \tag{4.44}$$

Assuming $p(L) = p_L$ in Eq. (4.44), the formula for the mass flow rate of gas takes a form similar to Eq. (4.40)

$$\dot{M} = \frac{\pi}{4}\sqrt{\frac{p_{0*}{}^2 - p_L{}^2}{\lambda \cdot Z_{av} R T_{av} \cdot L_*}} \cdot d^5 \qquad (4.45)$$

in which the initial pressure and the length of the pipeline section are changed to

$$\begin{cases} p_{0*} = p_0 \cdot \sqrt{\exp\left[\dfrac{2g}{Z_{av} R T_{av}}(z_0 - z_L)\right]}, \\ L_* = L \cdot \left(\dfrac{1}{L} \cdot \displaystyle\int_0^L \exp\left[\dfrac{2g}{Z_{av} R T_{av}}\left(z\left(\varsigma\right) - z_L\right)\right] \mathrm{d}\varsigma\right) \end{cases} \qquad (4.46)$$

It should be noted that the ratio $2g(z - z_L)/(Z_{av} R T_{av})$ is usually small when the difference in pipeline elevations is 100–1000 m ($Z_{av} R T_{av} \approx 150\,000$ m^2 s^{-1}). Therefore Eq. (4.46) could be simplified by expansion of the exponential function in the Taylor series. Accurate to the second term we obtain

$$\begin{cases} p_{0*} = p_0 \cdot \left(1 + \dfrac{g}{Z_{av} R T_{av}}(z_0 - z_L)\right), \\ L_* = L \cdot \left(1 + \dfrac{2g}{Z_{av} R T_{av}}(z_{av} - z_L)\right) \end{cases} \qquad (4.47)$$

Here z_{av} is the average elevation of the gas-pipeline section

$$z_{av} = \frac{1}{L} \cdot \int_0^L z(\varsigma)\, \mathrm{d}\varsigma.$$

Equations (4.47) shows that, even in the case when elevations of the beginning and the end of the pipeline section coincide, that is at $z_0 = z_L$, the length of the section has to be changed. At $z_{av} > z_L$, it should be enhanced (gas-pipelines running through a mountain pass), whereas at $z_{av} < z_L$ it should be diminished (gas-pipelines running along the sea bottom).

4.6.3 Calculation of Stationary Operating Regimes of a Gas-Pipeline (General Case)

To calculate the stationary operating regimes of a gas-pipeline the following system of ordinary differential equations is used

$$\begin{cases} \dfrac{\mathrm{d}p}{\mathrm{d}x} = -\lambda \dfrac{1}{d} \cdot \dfrac{\rho v^2}{2} - \rho g \cdot \sin\alpha(x), \quad \sin\alpha = \dfrac{\mathrm{d}z}{\mathrm{d}x}, \quad (v > 0), \\ \rho v C_p \dfrac{\mathrm{d}T}{\mathrm{d}x} - \rho v C_p D_* \dfrac{\mathrm{d}p}{\mathrm{d}x} = -\dfrac{4K}{d}(T - T_{ex}) \end{cases}$$

When taking into account that $\rho g\,\mathrm{d}z \ll \mathrm{d}p$ we have a system

$$\begin{cases} \dfrac{\mathrm{d}p}{\mathrm{d}x} = -\lambda\dfrac{1}{d}\cdot\dfrac{\rho v^2}{2}, \\ \rho v C_p\dfrac{\mathrm{d}T}{\mathrm{d}x} = -\dfrac{4K}{d}(T-T_{\mathrm{ex}}) - \rho v C_p D_* \cdot \lambda\dfrac{1}{d}\cdot\dfrac{\rho v^2}{2} \end{cases} \tag{4.48}$$

This system should be supplemented with the following equations

$$\rho v S = \dot{M},\ \rho = \frac{p}{(ZRT)},\quad Z = Z(p,T);$$

$$D_* = \frac{-RT(\partial Z/\partial p)_T}{C_p};\ \lambda = 0.067\cdot\left(\frac{2\Delta}{d}\right)^{0.2}$$

as well as with *initial* or *initial and boundary* conditions.

The following problems can be solved:

- the pressure p_0 and temperature T_0 at the initial cross-section of the gas-pipeline section as well as the mass flow rate $\dot{M}$ of gas are known. It is required to find the pressure p_L and temperature T_L at the end of the gas-pipeline section;
- the pressure p_0 and temperature T_0 at the initial cross-section of the gas-pipeline section as well as the pressure p_L at the end of the gas-pipeline section are known. It is required to find the mass flow rate $\dot{M}$ and the temperature T_L at the end of the gas-pipeline section;
- the pressure p_0 and temperature T_0 at the initial cross-section of the gas-pipeline section and the pressure p_L at the end of the gas-pipeline section as well as mass flow rate $\dot{M}$ of gas are known. It is required to find the diameter d of the gas-pipeline providing this flow rate;
- at the initial cross-section of the gas-pipeline section there is a *compressor station*. The pressure p_{en}, flow rate Q_{en} and temperature T_{en} of gas at the entrance to the compressor station and the pressure p_L at the end of the gas-pipeline section are known. It is required to choose the number and type of *gas-pumping aggregates*, compression ratio $\varepsilon = p_0/p_{\mathrm{en}}$ and the number of revolutions n of the centrifugal blower shafts.

4.6.4
Investigation of Thermal Regimes of a Gas-Pipeline Section

To consider the thermal regimes of gas transportation it is convenient to pass into the phase plane of variables (p, O). Dividing the second equation of the system (4.48) by the first equation of the same system, we get the differential equation

$$\frac{\mathrm{d}T}{\mathrm{d}p} = D_*(p,T) + \frac{8}{q_M^3}\cdot\frac{K}{\lambda}\cdot\frac{\rho(T-T_{\mathrm{ex}})}{C_p(p,T)} \tag{4.49}$$

containing only one unknown function $T(p)$. This equation should be solved at the segment $p_L \leq p \leq p_0$ under the condition $T(p_0) = T_0$, where p_0 and T_0 are the pressure and temperature, respectively, of the gas at the beginning of the pipeline section; p_L the pressure at the section end being unknown and to be determined; $q_M = Q_M S$ is the specific mass flow rate.

If we neglect the Joule–Thomson effect ($D_* \approx 0$), the right-hand side of Eq. (4.49) would be positive at $T > T_{ex}$ and negative at $T < T_{ex}$. Hence, in the first case the temperature of gas monotone increases with pressure whereas in the second case it monotone decreases. Mathematically speaking, the straight line $T = T_{ex}$ parallel to the abscissa axis is called the separatrix of Eq. (4.49) because it separates solutions of different types. When the initial temperature of the gas exceeds the temperature of the environment, the gas cools when moving from higher to lower pressure, whereas in the opposite case it heats.

The inclusion of the Joule–Thomson effect changes the pattern of solution of Eq. (4.49). On the plane (p, T) exists as before a separatrix separating the different types of solutions of Eq. (4.49). However, this separatrix is no longer a straight line. It always lies below the straight line $T = T_{ex}$, i.e. it appears to be in the temperature region below the temperature of the surrounding medium (see Fig. 4.5, curve 4).

If the temperature of the environment is greater than T_{ex}, then $\mathrm{d}T/\mathrm{d}p > 0$. This means that when the pressure falls from p_0 at the beginning of the pipeline section to p_L at the end of the section, it may become less than the

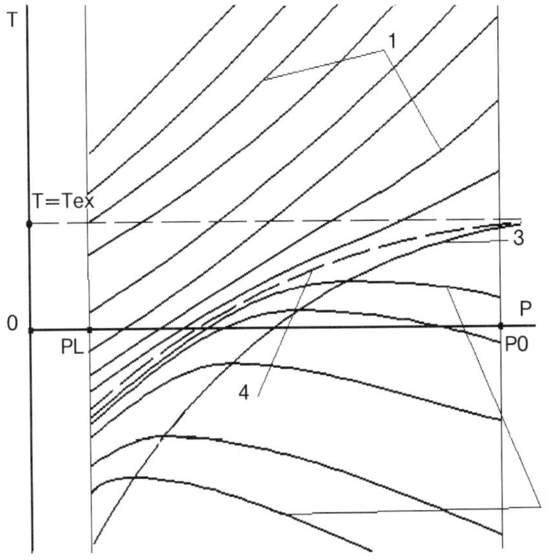

Figure 4.5 Solutions of differential equation (4.49): 1 – monotonic solutions; 2 – solutions with a maximum point; 3 – curve $\mathrm{d}T/\mathrm{d}p = 0$ of the temperature maxima; 4 – separatrix of solutions of Eq. (4.49).

temperature of the environment, since if $T = T_{ex}$ the derivative $dT/dp > 0$ is independent of the value of the heat-transfer factor (see Fig. 4.5 curve 1).

If the initial parameters of the gas are such that the point (p_0, T_0) is located below the separatrix 4 of Eq. (4.49), the derivative dT/dp could change sign (see Fig. 4.5, curves 2). The latter happens at points on line 3 determined by the condition of vanishing of the right-hand side of Eq. (4.49), namely

$$D_*(p, T) + \frac{8}{q_M^3} \cdot \frac{K}{\lambda} \cdot \frac{\rho(T - T_{ex})}{C_p(p, T)} = 0.$$

At points (p, T) of this curve $dT/dp = 0$, therefore the temperature of the gas at these points reaches an extremum, namely a maximum. The gas temperature first increases with decreasing pressure, at the point of intersection with curve 3 it reaches a maximum ($T_{max} < T_{ex}$) and then begins to diminish monotonically, remaining as before lower than the temperature of the environment.

The presence of maximum points on the curve $T(p)$ giving temperature distribution is related to processes having diverse actions: heating of the gas due to its mass exchange with the surrounding medium and cooling owing to the Joule–Thomson effect.

The phenomena under consideration may be used in exploitation of a gas pipeline section located in permafrost earth. If we provide an initial temperature of the gas T_0 such that the point (p_0, T_0) is located below the separatrix 4 of the solutions of Eq. (4.49), the temperature of the gas will remain below the temperature of the permafrost earth all the way along the pipeline section. This excludes the possibility of warming of the surrounding ground and imparts greater stability to the pipeline in the earth.

4.7
Modeling of Blower Operation

The motion of gas in a gas-pipeline is determined by compressor stations (GCS) located at the beginning of each pipeline section, or more precisely by blowers accomplishing gas compression. The main purpose of the blowers and of GCS as a whole is to force gas to move from a region of lower pressure (suction region at the GCS entrance) into a region of greater pressure (region of pumping at the GCS exit). The gas by itself cannot flow against pressure, therefore it is necessary to spend energy for forced flow in this direction. Such forced flow of gas against a pressure force is performed by gas-pumping aggregates (GPA) consisting of a drive (gas-turbine, electric, gas-motor and so on) producing rotation of an impeller shaft in centrifugal blowers (CFP) or the reciprocating motion of a piston in piston engines (PP), and by the blower itself. Displacement of gas from a region of lower pressure into a region of higher pressure, in other words gas *compression*, is accomplished in the blower. It is quite clear that gas in the gas-pipeline moves from a cross-section with greater pressure to a cross-section with lower pressure overcoming friction forces.

As a rule the compression station consists of separate plants equipped with several GPA with blowers connected in *parallel* in the case of single-step compression or in series in the case of *multi-step* compression. On main gas-pipelines centrifugal blowers are predominantly used following the same pattern as centrifugal blowers for fluid, see Fig. 4.1. Gas is sucked into the center of the impeller and thrown by centrifugal rotational force to the periphery of the impeller in the discharge line. The rotational rate of the impellers of a large CFP is 4000–15 000 rpm. The gas compression ratio depends on the type of blower, the rotational speed of its impeller (modern GPA have commonly controlled rotational speed), pressure and temperature at the discharge line and above all on gas flow rate.

A mathematical model of the centrifugal blower operating in the stationary regime involves algebraic dependences of the gas compression ratio ε and the developed specific power N/ρ_e on the gas parameters at the blower suction line indicated by the subscript e and the number of impeller revolutions n

$$\left[\begin{array}{l}\varepsilon = \dfrac{p_0}{p_B} = \varepsilon(\rho_B, p_B, T_B, Q_B, n, \ldots), \\ N = N(\rho_B, p_B, T_B, Q_B, n, \ldots)\end{array}\right. \tag{4.50}$$

where the dots denote parameters related to structural features of the blower.

From dimensional theory (see Chapter 6) it follows that these dependences may be represented as

$$\left[\begin{array}{l}\varepsilon = \dfrac{p_0}{p_B} = f\left(\dfrac{p_B/\rho_B}{n^2 D^2_{im}}, \dfrac{Q_B/\sigma}{n D_{im}}, \ldots\right), \\ \dfrac{N}{\rho_B} = \left(\dfrac{n}{n_0}\right)^3 \cdot \phi\left(\dfrac{p_B/\rho_B}{n^2 D^2_{im}}, \dfrac{Q_B/\sigma}{n D_{im}}, \ldots\right)\end{array}\right. \tag{4.51}$$

where D_{im}, σ, n_0 are the diameter of the impeller, the area of the suction branch pipe cross-section, the nominal number of revolutions of the blower shaft, respectively. If the structural parameters of the blower are taken as invariable dependences, Eq. (4.50) takes the following form

$$\left[\begin{array}{l}\varepsilon = \dfrac{p_0}{p_B} = f\left(\dfrac{p_B/\rho_B}{n^2}, \dfrac{Q_B}{n}, \ldots\right), \\ \dfrac{N}{\rho_B} = \left(\dfrac{n}{n_0}\right)^3 \cdot \phi\left(\dfrac{p_B/\rho_B}{n^2}, \dfrac{Q_B}{n}, \ldots\right)\end{array}\right. \quad \text{or}$$

$$\left[\begin{array}{l}\varepsilon = \dfrac{p_0}{p_B} = f\left(\dfrac{Z_B R T_B}{n^2}, \dfrac{Q_B}{n}, \ldots\right), \\ \dfrac{N}{\rho_B} = \left(\dfrac{n}{n_0}\right)^3 \cdot \phi\left(\dfrac{Z_B R T_B}{n^2}, \dfrac{Q_B}{n}, \ldots\right)\end{array}\right.$$

To determine the *universal* characteristic of the centrifugal blower the so-called *reduced* conditions, denoted by subscripts r, are used. The number of blower shaft revolutions in such cases is taken equal to a nominal value n_0, the properties of the gas and the conditions at the blower entrance are fixed: $R = R_r; T = T_r; Z = Z_r$. Tests of blowers conducted under these conditions permit the determination of the functions

$$\left[\begin{array}{l} \varepsilon = f\left(\dfrac{Z_r R_r T_r}{{n_0}^2}, \dfrac{Q_B}{n_0}, \ldots\right), \\ \left(\dfrac{N}{\rho}\right)_r = \phi\left(\dfrac{Z_r R_r T_r}{{n_0}^2}, \dfrac{Q_B}{n_0}, \ldots\right) \end{array}\right. \tag{4.52}$$

It is evident that the operating characteristics of a blower observed under *arbitrary conditions* but not under reduced conditions could be found from the universal characteristics written as follows

$$\left[\begin{array}{l} \varepsilon = f\left(\dfrac{Z_B R_B T_B}{n^2}, \dfrac{Q_B}{n}\right) = f\left(\dfrac{Z_r R_r T_r}{{n_0}^2} \cdot \dfrac{{n_0}^2}{n^2} \dfrac{Z_B R_B T_B}{Z_r R_r T_r}, \dfrac{Q_B \cdot n_0/n}{n_0}\right), \\ \dfrac{N}{\rho_B} = \left(\dfrac{n}{n_0}\right)^3 \cdot \left(\dfrac{N}{\rho_B}\right)_r \\ \quad = \left(\dfrac{n}{n_0}\right)^3 \cdot \phi\left(\dfrac{Z_r R_r T_r}{{n_0}^2} \cdot \dfrac{{n_0}^2}{n^2} \dfrac{Z_B R_B T_B}{Z_r R_r T_r}, \dfrac{Q_B \cdot n_0/n}{n_0}\right) \end{array}\right.$$

It follows that the operation characteristics of each blower under arbitrary conditions are obtained from some universal characteristics of the same blower, called *reduced*, through division of the arguments of these characteristic dependences governing the parameters by

$$\left(\frac{n}{n_0}\right)^2 \cdot \frac{Z_r R_r T_r}{Z_B R T_B} \quad \text{and} \quad \frac{n}{n_0},$$

respectively. This conclusion has a simple geometrical interpretation. If in the space to build the graphics of the dependences

$$\varepsilon = \tilde{f}[(n/n_0)_r, (Q_B)_r] \quad \text{and} \quad (N/\rho_B)_r = \tilde{\phi}[(n/n_0)_r, (Q_B)_r],$$

the characteristics of a centrifugal blower operating under arbitrary entrance conditions and at a number of revolutions n distinct from the nominal value n_0 are determined from these graphics by lengthening of the argument axes by factors $n/n_0 \cdot \sqrt{Z_r R_r T_r / Z_B R T_B}$ and n/n_0, respectively. At this the graphic of the second dependence is also stretched by the factor $(n/n_0)^3$ in the direction of the function axis.

If we introduce

$$\left(\frac{n}{n_0}\right)_r = \frac{n}{n_0} \cdot \sqrt{\frac{Z_r R_r T_r}{Z_B R T_B}} \quad \text{and} \quad (Q_B)_r = Q_B \cdot \frac{n_0}{n}. \tag{4.53}$$

The characteristics of the centrifugal blower take the universal form

$$\begin{bmatrix} \varepsilon = \tilde{f}\left[\left(\frac{n}{n_0}\right)_r, (Q_e)_r\right], \\ \left(\frac{N}{\rho_B}\right)_r = \tilde{\phi}\left[\left(\frac{n}{n_0}\right)_r, (Q_e)_r\right] \end{bmatrix} \tag{4.54}$$

where

$$\frac{N}{\rho_B} = \left(\frac{n}{n_0}\right)^3 \cdot \left(\frac{N}{\rho_e}\right)_r.$$

In Fig. 4.6 are depicted reduced (universal) characteristics of one of the centrifugal blowers 370-18-1 produced in Russia. The parameters of this blower are: $n_0 = 4800$ rpm; $T_r = 288$ K; $Z_r = 0.9$; $R_r = 490$ J kg^{-1} K^{-1}.

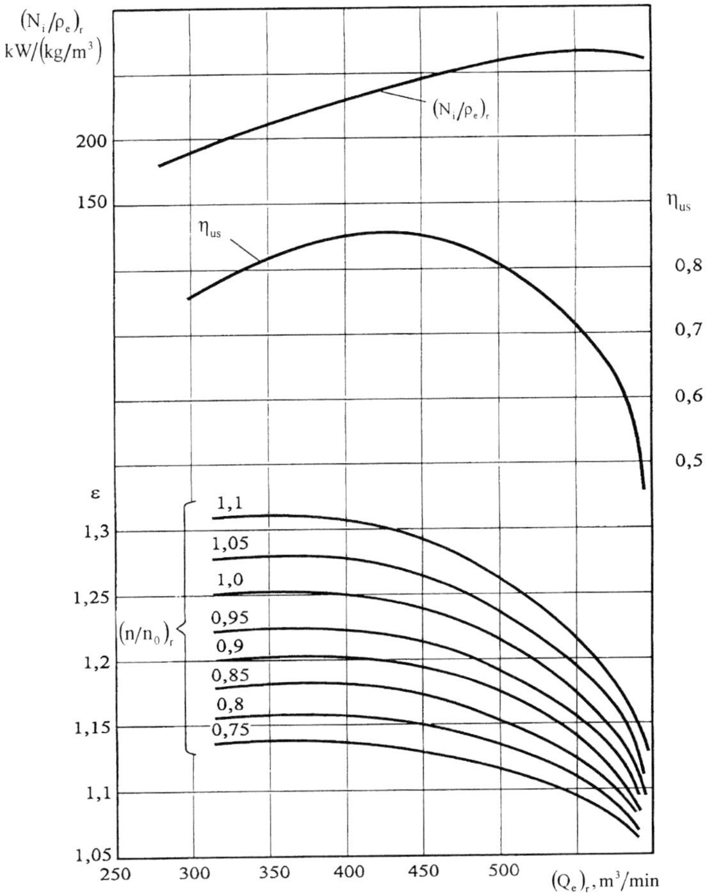

Figure 4.6 Reduced characteristics of the blower 370-18-1: $T_r = 288$ K; $Z_r = 0.9$; $R_r = 490$ J kg^{-1} K^{-1}.

Exercise. It is required to determine the rotational speed (number of revolutions per minute, rpm) of the centrifugal blower shaft 370-18-1 needed to provide transportation of natural gas ($\mu = 17.95$ kg kmol^{-1}, $p_{cr} = 4.7$ MPa, $T_{cr} = 194$ K) with commercial flow rate 22 million m^3 day^{-1} and compression ratio $\varepsilon = 1, 25$. It is known that the pressure and temperature of the gas at the suction line of the blower are 3.8 MPa and +15 °C, respectively.

Solution. Calculate first the parameters of the transported gas

$$R = \frac{8314}{17.95} \cong 463.1 \text{ J kg}^{-1}\text{ K}^{-1}),$$

with Eq. (2.14) we obtain:

$$Z_B = 1 - 0.4273 \cdot \frac{3.8}{4.7} \cdot \left(\frac{288}{194}\right)^{-3.668} \cong 0.919$$

$$\rho_{st} = \frac{p_{st}}{RT_{st}} = \frac{101\,300}{463.1 \cdot 293} \cong 0.746 \text{ kg m}^{-3}$$

$$\rho_B = \frac{p_B}{Z_B RT_B} = \frac{3.8 \cdot 10^6}{0.919 \cdot 463.1 \cdot 288} \cong 31.002 \text{ kg m}^{-3}$$

$$Q_B = Q_k \cdot \rho_{st}/\rho_B = \frac{22 \cdot 10^6}{24 \cdot 60} \cdot \frac{0.746}{31.002} \cong 367.6 \text{ m}^3\text{ min}^{-1}.$$

Then determine the reduced parameters of the operating regime of the centrifugal blower

$$\left(\frac{n}{n_0}\right)_r = \frac{n}{n_0}\sqrt{\frac{Z_r R_r T_r}{Z_B RT_B}} = \frac{n}{n_0}\sqrt{\frac{0.90 \cdot 490 \cdot 288}{0.919 \cdot 463.1 \cdot 288}} \cong 1.018 \cdot \frac{n}{n_0};$$

$$(Q_B)_r = Q_B\frac{n_0}{n} = 367.6 \cdot \frac{n_0}{n}\text{m}^3\text{ min}^{-1}.$$

Since the compression ratio ε is known and is equal to 1.25, it is necessary, using the reduced characteristics of the blower 370-18-1 (see Fig. 4.6), to select a value of n/n_0 such that the point with coordinates $Q_B = 367.6/(n/n_0)$ and $\varepsilon = 1.25$ would lie on the characteristic $(n/n_0)_r = 1.018 \cdot n/n_0$. The solution is sought by the iteration method.

1. Let $(n/n_0)_r = 0.9 \Rightarrow n/n_0 \cong 0.916$;
   $(Q_B)_r = 367.6/0.916 \cong 401$ m^3 min^{-1} $\Rightarrow \varepsilon \cong 1.2 < 1.25$ (see Fig. 4.6), consequently $(n/n_0)_r$ should be increased.
2. Let $(n/n_0)_r = 1.0 \Rightarrow n/n_0 = 1.0/1.018 \cong 0.982$;
   $(Q_B)_r = 367.6/0.982 \cong 374$ m^3 min^{-1} $\Rightarrow \varepsilon \cong 1.25$ (see Fig. 4.6), consequently the solution can be taken as correct.
   Hence $n = 0.982 \cdot n_0 = 0.982 \cdot 4800 \cong 4714$ rpm.

Answer. 4714. rpm.

Useful Power of a Blower

Let us show now how we can estimate the useful power needed for gas compression from pressure p_B at the blower entrance to pressure p_0 at the blower exit.

From the total energy balance equation (1.35) written for a mass of gas going between the entrance cross-section x_1 and the exit cross-section x_2 of a blower in the case of stationary flow it follows that

$$\left[\left(\frac{\alpha_k v^2}{2} + e_{in} + \frac{p}{\rho}\right) \cdot \rho v S\right]_{x_2} - \left[\left(\frac{\alpha_k v^2}{2} + e_{in} + \frac{p}{\rho}\right) \cdot \rho v S\right]_{x_1} = q^{ex} + N_{us}$$

where q^{ex} is the external heat inflow ($q^{ex} > 0$) to the gas or heat outflow from the gas ($q^{ex} < 0$); N_{us} is the useful power of mechanical forces acting on the gas, i.e. the useful power of the blower. Taking into account that the mass flow rate of gas $\dot{M} = \rho v S$ through all gas-pipeline cross-sections in stationary flow remains constant, we have

$$\left[\left(\frac{\alpha_K {v_0}^2}{2} - \frac{\alpha_K {v_B}^2}{2} + (J_0 - J_B)\right)\right] \cdot \dot{M} = q^{ex} + N_{us}$$

where $J = e_{in} + p/\rho$ is the gas enthalpy. Neglecting the difference $(\alpha_K {v_0}^2 - \alpha_K {v_B}^2)/2$ in the kinetic energies of the gas before and after compression and assuming the process of gas compression in the blower to be adiabatic ($q^{ex} = 0$), we get

$$N_{us} \cong \dot{M} \cdot (J_0 - J_B) = \rho_B Q_B \cdot (J_0 - J_B) \tag{4.55}$$

where $Q = vS$ is the volume flow rate of gas ($\rho_B Q_B = \rho_0 Q_0 = \text{const.}$).

For a perfect gas $J = C_p T + \text{const.}$ and the following relations are valid

$$N_{us} = \rho_B Q_B \cdot (J_0 - J_B) = \rho_B C_p Q_B \cdot (T_0 - T_B) = \frac{C_p}{R} \frac{p_B Q_B}{T_B} (T_0 - T_B)$$

or

$$N_{us} = \frac{C_p}{C_p - C_v} \cdot p_B Q_B \left(\frac{T_0}{T_B} - 1\right) = \frac{\gamma}{\gamma - 1} \cdot p_B Q_B \left(\frac{T_0}{T_B} - 1\right)$$

where $\gamma = C_p/C_v$ is the adiabatic index (for methane $\gamma \approx 1.31$).

Taking into account that in an adiabatic process the temperature T varies in accordance with the power law $T/T_e = (p/p_e)^{\frac{\gamma-1}{\gamma}}$, we finally obtain

$$N_{us} = \frac{\gamma}{\gamma - 1} \cdot p_B Q_B \left(\varepsilon^{\frac{\gamma-1}{\gamma}} - 1\right). \tag{4.56}$$

Here $\varepsilon = p_0/p_B$ is the compression ratio.

A similar formula written as a function of the gas parameters at the exit of the blower has the form

$$N_{us} = \frac{\gamma}{\gamma - 1} \cdot p_0 Q_0 \left(1 - \varepsilon^{\frac{1-\gamma}{\gamma}}\right). \tag{4.57}$$

Here the subscript 0 denotes that the relevant quantity is taken at the exit of the blower.

Exercise. The pressure p_B before the blower is 3.5 MPa, the compression ratio of a gas with $\gamma = 1.31$ is 1.4, the volume flow rate at the entrance Q_B of the blower is 500 m^3 min^{-1}. It is required to determine the useful power N_{us} spent by the blower for gas compression.

Solution. The useful power is obtained from Eq. (4.48)

$$N_{us} = \frac{1.31}{1.31 - 1} \cdot 3.5 \cdot 10^6 \cdot \frac{500}{60} \cdot \left(1.4^{\frac{1.31-1}{1.31}} - 1\right) \cong 10.2 \cdot 10^6 \text{ W}.$$

Hence the useful power is 10.2 MW.

Answer. 10.2 MW.

Finally in this chapter let us consider an exercise on the calculation of a gas-pipeline section in combination with a compressor station.

Exercise. Natural gas ($R = 18.82$ kg kmol^{-1}, $p_{cr} = 4.75$ MPa, $T_{cr} = 195$ K) is being transported along a 105-km gas-pipeline section ($D = 1220 \times 12$ mm, $\Delta = 0.03$ mm) with the help of two identical GPA equipped with blowers 370-18-1 connected in parallel. It is required to determine the compression ratio of the gas ε and the number of revolutions n of the blower rotors needed to ensure a commercial flow rate of 21 billion m^3 year^{-1} in the gas-pipeline (the number of working days in a year is taken to be 350). It is given that the pressure at the end of the pipeline section is 3.8 MPa and at the blower suction line is 4.7 MPa; the temperature of the gas at the suction line is 12 °C; the temperature of the gas after compression is expected to be 30 °C; the temperature of the surrounding ground is 8 °C.

Solution. Taking in Eq. (4.33) $x = L$ and $Q_k = \dot{M}/\rho_{st}$, we have

$$p_0{}^2 = p_L{}^2 + \frac{16\lambda \cdot Z_{av} R T_{av} \rho_{st}{}^2 \cdot L}{\pi^2 d^5} \cdot Q_k{}^2.$$

Now calculate successively the quantities in this relation

$$Q_k = \frac{21 \cdot 10^9}{350 \cdot 24 \cdot 3600} \cong 694.4 \text{ m}^3 \text{ s}^{-1};$$

$$R = \frac{R_0}{\mu} = \frac{8314}{18.82} \cong 441.7 \text{ J kg}^{-1} \text{ K}^{-1};$$

$$\rho_{st} = \frac{p_{st}}{RT_{et}} = \frac{10\,125}{441.7 \cdot 293} \cong 0.783 \text{ kg m}^{-3};$$

$$\lambda = 0.067 \cdot \left(\frac{2\Delta}{d}\right)^{0.2} = 0.067 \cdot \left(\frac{2 \cdot 0.03}{1196}\right)^{0.2} \cong 0.0093.$$

Calculate the average gas temperature T_{av} in the pipeline section

$$T_{av} = T_{ex} + \frac{T_0 - T_L}{\ln\left(\frac{T_{0.} - T_{ex}}{T_L - T_{ex}}\right)} = 8 + \frac{30 - 12}{\ln\left(\frac{30 - 8}{12 - 8}\right)} \cong 18.6\,^{\circ}\text{C} = 291.6 \text{ K}.$$

The gas over-compressibility factor Z is calculated with Eq. (2.16) taking the average pressure p_{av} in a first approximation to be equal to the pressure at the end of the pipeline section and the temperature to be the temperature averaged over the section

$$Z = 1 - 0.4273 \cdot (3.8/4.75) \cdot (291.6/195)^{-3.668} \cong 0.922.$$

After this we calculate the pressure p_0 at the beginning of the pipeline section. We begin with the factor A

$$A = \frac{16 \cdot \lambda \cdot Z_{av} R T_{av} \rho_{st}{}^2 \cdot L}{\pi^2 d^5} \cdot Q_k{}^2$$

$$= \frac{16 \cdot 0.0093 \cdot 0.922 \cdot 441.7 \cdot 291.6 \cdot 0.783^2 \cdot 105\,000}{3.14^2 \cdot 1.196^5} \cdot 694.4^2$$

$$\cong 22.73 \cdot 10^{12}$$

and then calculate the pressure p_0

$$p_0 = \sqrt{(3.8 \cdot 10^6)^2 + 22.73 \cdot 10^{12}} \cong 6.1 \cdot 10^6 \text{ Pa or } 6.1 \text{ MPa}.$$

The obtained value shows that the average pressure p_{av} in the pipeline section is equal to $2/3 \cdot (4.5 + 6.1^2/10.6) \cong 5.34$ MPa, which is greater than the expected 3.8 MPa. Hence, the calculation should be corrected.

Performing the second approximation for pressure $p = p_{av} = 5.34$ MPa, we get

$$Z = 1 - 0.4273 \cdot (5.34/4.75) \cdot (291.6/195)^{-3.668} \cong 0.890;$$

$$A \cong 21.94 \cdot 10^{12} \text{ Pa}^2;$$

$$p_0 = \sqrt{(3.8 \cdot 10^6)^2 + 21.94 \cdot 10^{12}} \cong 6.0 \cdot 10^6 \text{ Pa or } 6.0 \text{ MPa}.$$

We see that the obtained value of p_0 is practically unchanged. Thus the compression ratio ε, which should be provided by the blowers 370-18-1, is $6.0/4.7 \cong 1.28$.

Once the required compression ratio has been obtained, it is time to calculate the gas parameters at the suction line of each of the blowers connected inparallel. We have

$$Z_B = 1 - 0.4273 \cdot \frac{4.7}{4.75} \cdot \left(\frac{285}{195}\right)^{-3.668} \cong 0.895;$$

$$\rho_B = \frac{p_B}{Z_B R T_B} = \frac{4.7 \cdot 10^6}{0.895 \cdot 441.7 \cdot 285} \cong 41.716\ \text{kg m}^{-3}.$$

In a parallel connection of identical blowers the flow rate is distributed equally between them, therefore it is

$$Q_B = Q_k \cdot \rho_{st}/\rho_B = \frac{[(21\,000/2)/350] \cdot 10^6}{24 \cdot 60} \cdot \frac{0.783}{41.716} \cong 391\ \text{m}^3\ \text{min}^{-1}.$$

Determine the reduced parameters of the operating regime of the centrifugal blower:

$$\left(\frac{n}{n_0}\right)_r = \frac{n}{n_0}\sqrt{\frac{Z_r R_r T_r}{Z_B R T_B}} = \frac{n}{n_0}\sqrt{\frac{0.90 \cdot 490 \cdot 288}{0.895 \cdot 441.7 \cdot 285}} \cong 1.062 \cdot \frac{n}{n_0};$$

$$(Q_B)_r = Q_B \frac{n_0}{n} = 391 \cdot \frac{n_0}{n}\ \text{m}^3\ \text{min}^{-1}.$$

As the compression ratio ε has already been found to be equal to 1.28, it is necessary, using the characteristics of the blower 370-18-1 depicted in Fig. 4.6, to select n/n_0 in such a way that a point with coordinates $(Q_B)_r = 391/(n/n_0)$ and $\varepsilon = 1.28$ would lie on the characteristic $(n/n_0)_r = 1.062 \cdot n/n_0$. The selection is performed by the iteration method.

1. We take $(n/n_0)_r = 1.0 \Rightarrow n/n_0 = 1.0/1.062 \cong 0.942$;
   $(Q_B)_r = 391/0.942 \cong 415\ \text{m}^3\ \text{min}^{-1} \Rightarrow \varepsilon \cong 1.25$ (see Fig. 4.6), which is less than the required value 1.28. Therefore $(n/n_0)_r$ should be increased.
2. Let now $(n/n_0)_r = 1.05 \Rightarrow n/n_0 = 1.05/1.062 \cong 0.989$;
   $(Q_B)_r = 391/0.989 \cong 395\ \text{m}^3\ \text{min}^{-1} \Rightarrow \varepsilon \cong 1.28$ (see Fig. 4.6). Hence, the solution is found.
   As a result we have $n = 0.989 \cdot n_0 = 0.989 \cdot 4800 \cong 4750$ rpm.

5
Closed Mathematical Models of One-Dimensional Non-Stationary Flows of Fluid and Gas in a Pipeline

In this chapter are considered the most important models of one-dimensional non-stationary flows of fluid and gas in pipelines. The equations obtained in Chapter 1 are used as a starting point.

5.1
A Model of Non-Stationary Isothermal Flow of a Slightly Compressible Fluid in a Pipeline

At the basis of this model lie the following assumptions:

- the variation of fluid density $\Delta\rho$ is much less than its nominal value ρ_0, that is $\Delta\rho \ll \rho_0$, where $\Delta\rho = \rho_0 \times (p - p_0)/K$ in accordance with Eq. (2.6). For example, at $\rho_0 = 1000$ kg m^{-3}, $p - p_0 = 1.0$ MPa (10^6 Pa ≈ 10 atm), $K = 10^3$ MPa (10^9 Pa), the variation of fluid density $\Delta\rho$ is only 1 kg m^{-3};
- the variation of pipeline cross-section area ΔS is much less than its nominal value S_0, that is $\Delta S \ll S_0$, where $\Delta S = \pi {d_0}^3/4E\delta \cdot (p - p_0)$ or $\Delta S = S_0 d_0/E\delta \cdot (p - p_0)$. For example, at $d_0 = 500$ mm, $\delta = 10$ mm, $\rho_0 = 10^3$ kg m^{-3}, $p - p_0 = 10^7$ Pa ($\approx$100 atm), $E = 2 \cdot 10^{11}$ Pa (pipe steel), the variation of the pipeline diameter Δd is 0.06 mm, the variation of the pipeline cross-section area ΔS is $\approx$0.5 cm^2, whereas $S_0 \cong 1960$ cm^2;
- the tangential friction stress $|\tau_w|$ at the pipeline walls in accordance with Eq. (1.28) is determined by the formula $|\tau_w| = \lambda(Re, \varepsilon) \cdot \rho v^2/8$ with the factor λ dependent on the governing parameters $Re = vd/\nu$ and $\varepsilon = \Delta/d$ given in the same form as in stationary flow. Such an assumption is called the *hypothesis of quasi-stationarity*. For example, for $\lambda = 0.02$, $v = 1.5$ m s^{-1}, $\rho_0 = 1000$ kg m^{-3}, then $|\tau_w| \cong 5.6$ Pa.

The first equation of the model is continuity equation

$$\frac{\partial \rho S}{\partial t} + \frac{\partial \rho v S}{\partial x} = 0.$$

With regard to the accepted assumptions this equation could be transformed to the following form

$$\frac{\partial \rho S}{\partial t} = \rho\frac{\partial S}{\partial t} + S\frac{\partial \rho}{\partial t} \approx \rho_0 \frac{dS}{dp}\frac{\partial p}{\partial t} + S_0\frac{d\rho}{dp}\frac{\partial p}{\partial t} = \left(\frac{\rho_0 S_0 d_0}{E\delta} + \frac{\rho_0 S_0}{K}\right)\cdot\frac{\partial p}{\partial t};$$

Modeling of Oil Product and Gas Pipeline Transportation. Michael V. Lurie

ISBN: 978-3-527-40833-7

$$\frac{\partial \rho v S}{\partial x} \approx \rho_0 S_0 \cdot \frac{\partial v}{\partial x}.$$

As a result the following equation is obtained

$$\left(\frac{\rho_0}{K} + \frac{\rho_0 d_0}{E\delta}\right) \cdot \frac{\partial p}{\partial t} + \rho_0 \frac{\partial v}{\partial x} = 0.$$

The factor in parentheses before the derivative of pressure with respect to time has dimension inverse to the square of the velocity, therefore it can be denoted as $1/c^2$, where the parameter c, given by

$$c = \frac{1}{\sqrt{\dfrac{\rho_0}{K} + \dfrac{\rho_0 d_0}{E\delta}}} \tag{5.1}$$

is called the *speed of wave propagation in the pipeline* ($c \approx 1000$ m s^{-1}). If $\rho_0 = 1000$ kg m^{-3}, $K = 10^9$ Pa, $d_0 = 500$ mm, $\delta = 10$ mm, $E = 2 \cdot 10^{11}$ Pa, then

$$c = \frac{1}{\sqrt{\dfrac{10^3}{10^9} + \dfrac{10^3 \cdot 0.5}{2 \cdot 10^{11} \cdot 0.01}}} \cong 895 \text{ m s}^{-1}.$$

With regard to the introduced designation *the first equation of the model* takes the form

$$\frac{\partial p}{\partial t} + \rho_0 c^2 \cdot \frac{\partial v}{\partial x} = 0. \tag{5.2}$$

The second equation of the model is the momentum equation (1.10)

$$\rho \left(\frac{\partial v}{\partial t} + v \frac{\partial v}{\partial x}\right) = -\frac{\partial p}{\partial x} - \frac{4}{d_0} \tau_w - \rho g \sin \alpha(x)$$

Replacement of τ_w with the expression containing the average velocity of the flow v yields

$$\rho \left(\frac{\partial v}{\partial t} + v \frac{\partial v}{\partial x}\right) = -\frac{\partial p}{\partial x} - \lambda \frac{1}{d_0} \frac{\rho v|v|}{2} - \rho g \sin \alpha(x). \tag{5.3}$$

For slightly compressible fluids, among which are water, oil and oil products, the following simplifying assumptions can be made

$$\frac{\partial v}{\partial t} \approx \rho_0 \frac{\partial v}{\partial t},$$

$$\rho v \frac{\partial v}{\partial x} \cong \rho_0 v \frac{\partial v}{\partial x} = \frac{\partial}{\partial x}\left(\frac{\rho_0 v^2}{2}\right),$$

$$\frac{\partial}{\partial x}\left(p + \frac{\rho_0 v^2}{2}\right) \approx \frac{\partial p}{\partial x}.$$

The last approximation is valid because it is easy to verify that $\Delta(\rho_0 v^2/2) \ll \Delta p$. Really, at $\rho_0 \approx 1000\ \mathrm{kg\ m^{-3}}$ and $v \approx 1\text{–}2\ \mathrm{m\ s^{-1}}$, $\Delta(\rho_0 v^2/2) \leq 2000$ Pa ($\approx$0.04 atm), whereas Δp is measured in atmospheres or even tens of atmospheres. In the general case $\Delta(\rho_0 v^2/2) \approx \rho_0 v\ \Delta v$ while $\Delta p \approx \rho_0 c\ \Delta v$, hence $\Delta(\rho_0 v^2/2)/\Delta p \approx v/c$. Since $v \ll c$ in pipelines, the ratio $\Delta(\rho_0 v^2/2)/\Delta p$ is negligibly small.

With regard to the given assumptions the momentum equation takes the following form

$$\rho_0 \frac{\partial v}{\partial t} + \frac{\partial p}{\partial x} = -\lambda(Re, \varepsilon) \cdot \frac{1}{d_0} \frac{\rho_0 v|v|}{2} - \rho_0 g \sin\alpha(x)$$

This equation is the *second equation of the model.*

Hence, the mathematical model of slightly compressible fluids is represented by a system of two differential equations

$$\begin{cases} \dfrac{\partial p}{\partial t} + \rho_0 c^2 \cdot \dfrac{\partial v}{\partial x} = 0, \\ \rho_0 \dfrac{\partial v}{\partial t} + \dfrac{\partial p}{\partial x} = -\lambda(Re, \varepsilon) \dfrac{1}{d_0} \dfrac{\rho_0 v|v|}{2} - \rho_0 g \sin\alpha(x) \end{cases} \tag{5.4}$$

to determine the two unknown functions $p(x, t)$ and $v(x, t)$ dependent on the coordinate x and time t.

The system of differential equations (5.4) requires for its solution *initial and boundary conditions,* being also components of the model under consideration. We will deal with these conditions below.

Virtual Mass

The hypothesis of quasi-stationarity in accordance with which the tangential stress τ_w at the internal surface of the pipe is represented by the equality $\tau_w = \lambda(Re, \varepsilon) \cdot \rho_0 v|v|/8$, asserts in particular that it depends on the *instantaneous value* of the average flow velocity $v(x, t)$ but not on derivatives of the velocities with respect to time and coordinate. In fact the following equation holds

$$\frac{4}{d}\tau_w \cdot vS = \rho_0 g v S \cdot i_0 + \rho_0 S \frac{\mathrm{d}}{\mathrm{d}t}\left[(\alpha_k - 1)\frac{v^2}{2}\right] \tag{5.5}$$

(see Eq. (1.23)), reflecting the transformation of the work of friction forces ($4\tau_w \cdot vS/d$), and in one-dimensional flow appearing as external forces, into kinetic energy $\rho_0 S \frac{\mathrm{d}}{\mathrm{d}t}[(\alpha_k - 1)\frac{v^2}{2}]$ of the *intrinsic* motion of the fluid layers relative to its center of mass and into heat $\rho_0 g S(v \cdot i_0)$ due to the work of internal friction forces. The values of the factor α_k in this equation vary from 4/3 for laminar flow up to 1.02–1.05 for turbulent flow. If we take for α_k the mean value of this factor, from Eq. (5.5) we get

$$\frac{4}{d}\tau_w = \rho_0(\alpha_k - 1) \cdot \frac{\mathrm{d}v}{\mathrm{d}t} + \rho_0 g \cdot i_0 \tag{5.6}$$

indicating that the tangential stress τ_w contains the term $\rho_0(\alpha_k - 1) \cdot dv/dt$ proportional to the fluid particle acceleration. The physical nature of this term is hidden in the origin of the additional resistance force $\rho_0(\alpha_k - 1) \cdot \dot{v}$ caused by realignment of the internal structure of the flow taking place even when there is dissipation of mechanical energy into heat owing to the work of the viscous friction force, that is the term $\rho_0 g \cdot i_0$, could be neglected. It is evident that if $v = \text{const.}$ the first term vanishes.

Substitution of Eq. (5.6) into the momentum equation (5.2) yields

$$\rho_0 \frac{dv}{dt} = -\frac{\partial p}{\partial x} - \rho_0(\alpha_k - 1)\frac{dv}{dt} - \rho_0 g i_0 - \rho_0 g \sin\alpha(x)$$

or

$$\alpha_k \rho_0 \frac{dv}{dt} = -\frac{\partial p}{\partial x} - \lambda(Re, \varepsilon) \cdot \frac{1}{d_0} \frac{\rho_0 v|v|}{2} - \rho_0 g \sin\alpha(x) \tag{5.7}$$

Equation (5.7) differs from Eq. (5.3) only in that in the left-hand side enters not the true fluid density ρ_0 but a quantity $\alpha_k \rho_0$ differing from ρ_0 by the factor α_k. The quantity $\rho_0(\alpha_k - 1)$ may be called the *virtual (additional) mass* of the fluid. Hence, the inertial properties of a fluid in non-stationary processes are characterized by changing density, to the latter is added a certain quantity dependent on the flow regime. In developed turbulent flows this change is slight ($\alpha_k \approx 1.03$), but for laminar flow it is greater ($\alpha_k \approx 4/3$).

5.2 A Model of Non-Stationary Gas Flow in a Pipeline

At the basis of this model lie the following assumptions:

- the transported media (gas) is compressible, i.e. $\rho = \rho(p, T)$;
- the variation of gas-pipeline cross-section area ΔS can be ignored compared with the area itself S_0, i.e. $\Delta S \ll S_0$. Therefore $S \cong S_0 = \pi {d_0}^2/4 = \text{const.}$;
- the internal energy of the gas is $e_{in} = C_v T + \text{const.}$

The *system of basic equations* is

$$\begin{cases} \dfrac{\partial \rho}{\partial t} + \dfrac{\partial \rho v}{\partial x} = 0, \\ \dfrac{\partial \rho v}{\partial t} + \dfrac{\partial}{\partial x}(p + \rho v^2) = -\dfrac{4\tau_w}{d_0} - \rho g \sin\alpha, \\ \rho\left(\dfrac{\partial e_{in}}{\partial t} + v\dfrac{\partial e_{in}}{\partial x}\right) = \dfrac{4 q_n}{d_0} - p\dfrac{\partial v}{\partial x} - \rho \cdot n_{in} \end{cases} \tag{5.8}$$

The *first equation* of this system is the continuity equation reflecting the law of gas mass conservation in each pipeline cross-section.

The *second equation* of the system is the momentum equation expressing Newton's second law.

The *third equation* of the system is the equation of heat inflow following from the laws of total energy conservation of the flow and the variation of mechanical energy of the transported media.

To the system of equations (5.8) should be added the so-called *closing relations,* e.g.

$$\tau_w = \frac{\lambda(Re, \varepsilon)}{8} \cdot \rho v^2, \quad n_{in} = -\lambda(Re, \varepsilon) \cdot \frac{1}{d_0} \frac{\rho v^3}{2},$$

$$\rho = \frac{p}{Z(p, T) \cdot RT}, \quad e_{in}(T) = C_v \cdot T + \text{const.}, \quad q_n = -\kappa \cdot (T - T_{ex}).$$

Hence, if $\lambda(Re, \varepsilon)$ and $Z(p, T)$ are known as functions of their arguments, the system of equations (5.8) represents a closed system of three partial differential equations for three unknown functions $p(x, t)$, $v(x, t)$ and $T(x, t)$ dependent on the coordinate x and time t.

5.3 Non-Stationary Flow of a Slightly Compressible Fluid in a Pipeline

Consider the non-stationary flow of a *slightly compressible fluid* in a pipeline. The basic equations of such flow are represented by the system of equations (5.4).

5.3.1 Wave Equation

Let us consider first the non-stationary flow of a slightly compressible fluid in a horizontal pipeline ($\alpha = 0$) neglecting for a while terms accounting for the friction force. Such an assumption is quite allowable for short pipelines and fluids with not too high viscosity. In such a case the system of equations (5.4) takes the form

$$\begin{cases} \dfrac{\partial p}{\partial t} + \rho_0 c^2 \cdot \dfrac{\partial v}{\partial x} = 0, \\ \rho_0 \dfrac{\partial v}{\partial t} + \dfrac{\partial p}{\partial x} = 0 \end{cases}$$

or

$$\begin{cases} \dfrac{\partial v}{\partial x} = -\dfrac{1}{\rho_0 c^2} \cdot \dfrac{\partial p}{\partial t}, \\ \dfrac{\partial v}{\partial t} = -\dfrac{1}{\rho_0} \cdot \dfrac{\partial p}{\partial x}. \end{cases} \tag{5.9}$$

Differentiation of the first equation with respect to t and the second with respect to x subject to the condition $v''_{x,t} = v''_{t,x}$ yields the equation for $p(x, t)$

$$\frac{\partial^2 p}{\partial t^2} = c^2 \frac{\partial^2 p}{\partial x^2}. \tag{5.10}$$

This equation is called the *wave equation* since it describes the propagation of waves and is encountered in different fields of physics.

A similar equation can be obtained for fluid velocity $v(x, t)$

$$\frac{\partial^2 v}{\partial t^2} = c^2 \frac{\partial^2 v}{\partial x^2}. \tag{5.11}$$

Equations (5.10) and (5.11) represent partial differential equations whose general solution is expressed by two arbitrary functions

$$p(x, t) = f_1(x - ct) + f_2(x + ct) \tag{5.12}$$

the first of these functions is dependent only on the argument $\xi = x - ct$ and the second on $\eta = x + ct$.

Let us show that Eq. (5.12) gives a solution of Eq. (5.10). We have

$$\frac{\partial p}{\partial t} = -cf'_{1\xi} + cf'_{2\eta},$$

$$\frac{\partial^2 p}{\partial t^2} = c^2 f''_{1\xi\xi} + c^2 f''_{2\eta\eta} = c^2 (f''_{1\xi\xi} + f''_{2\eta\eta}),$$

$$\frac{\partial p}{\partial x} = f'_{1\xi} + f'_{2\eta}; \quad \frac{\partial^2 p}{\partial x^2} = f''_{1\xi\xi} + f''_{2\eta\eta}.$$

$$c^2 \frac{\partial^2 p}{\partial x^2} = \frac{\partial^2 p}{\partial t^2}.$$

Hence, Eq. (5.12) is a solution of Eq. (5.10) for arbitrary functions f_1 and f_2.

The function $f_1(x - ct)$ represents a *traveling wave* in the positive direction of the x-axis whereas the function $f_2(x + ct)$ represents a traveling wave in the negative direction of this axis. The magnitude of the velocity propagation of both waves, i.e. the propagation velocity of a certain value of function f_1 or f_2 is identical and equal to c.

The form of each of the functions f_1 and f_2 is determined by the initial conditions for pressure and velocity distributions in the pipeline, that is by $p(x, 0)$ and $v(x, 0)$, as well as by the boundary conditions at the pipeline ends.

The velocity of fluid $v(x, t)$ is determined by the formula

$$v(x, t) = g_1(x - ct) + g_2(x + ct).$$

Functions $g_1(\xi)$ and $g_2(\eta)$ are expressed through functions $f_1(\xi)$ and $f_2(\eta)$. From Eq. (5.9) follows

$$\begin{cases} g'_{1\xi} + g'_{2\eta} = \dfrac{1}{\rho_0 c} \cdot [f'_{1\xi} - f'_{2\eta}], \\ -c \cdot g'_{1\xi} + c \cdot g'_{2\eta} = -\dfrac{1}{\rho_0}[f'_{1\xi} + f'_{2\eta}], \end{cases}$$

or

$$\begin{cases} \rho_0 c g'_{1\xi} = f'_{1\xi}, \\ \rho_0 c g'_{2\eta} = -f'_{2\eta}. \end{cases}$$

Integration of the first equation with respect to ξ and the second with respect to η, taking into account that g_1 and f_1 depend only on ξ and g_2 and f_2 only on η, yields

$$\begin{cases} \rho_0 c g_1 = f_1 + \text{const.}, \\ \rho_0 c g_2 = -f_2 + \text{const.} \end{cases}$$

From here follows

$$\rho_0 c \cdot v(x, t) = f_1(x - ct) - f_2(x + ct) = \text{const.} \tag{5.13}$$

With the help of Eqs. (5.12) and (5.13) one can find solutions to different problems. We will consider some of them.

5.3.2 Propagation of Waves in an Infinite Pipeline

A pipeline is called *infinite* when it runs in the direction of the x-axis from $-\infty$ to $+\infty$. Of course it is only a model of a real pipeline, but it is very useful in the case of a very long pipeline when boundary effects can be ignored, that is when waves reflected from the beginning and the end of a pipeline could be neglected. Suppose also that friction forces are absent.

Addition and subtraction of Eqs. (5.12) and (5.13) yield

$$\begin{cases} p + \rho_0 c \cdot v = 2 \cdot f_1(x - ct) + \text{const.}, \\ p - \rho_0 c \cdot v = 2 \cdot f_2(x + ct) + \text{const.} \end{cases}$$

From these expressions it is seen that at those points of the plane (x, t) where $(x - ct)$ remains constant the expression $I_1 = p + \rho_0 c \cdot v$ is also constant and at those points of the plane (x, t) where $(x + ct)$ remains constant the expression $I_2 = p - \rho_0 c \cdot v$ is also constant (see Fig. 5.1).

The lines $x - ct = \text{const.}$ and $x + ct = \text{const.}$ are called *characteristics of the wave equation* and the quantities $I_1 = p(x, t) + \rho_0 c \cdot v(x, t)$ and $I_2 = p(x, t) - \rho_0 c \cdot v(x, t)$ are called the *Riemann invariants*.

Hence, at each characteristic $x - ct = \text{const.}$ with positive slope, $\mathrm{d}x/\mathrm{d}t = +c$, the first Riemann invariant I_1 is conserved whereas at each characteristic

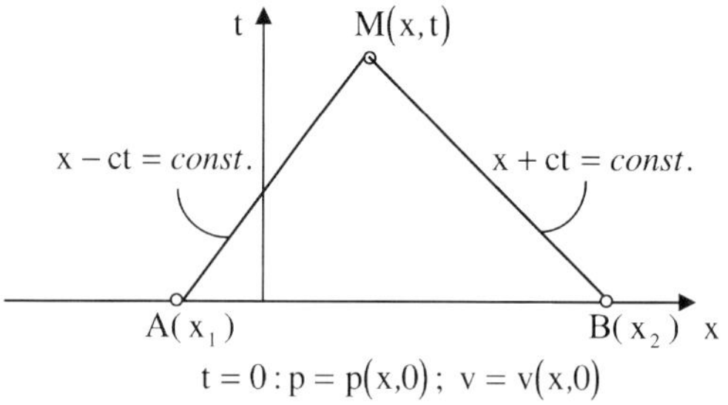

Figure 5.1 Characteristics on (x, t)-plane.

$x + ct =$ const. with negative slope, $dx/dt = -c$, the second Riemann invariant I_2 is conserved:

$$\text{at lines } x = ct + \text{const.} : I_1 = p + \rho_0 c v = \text{const.};$$

$$\text{at lines } x = -ct + \text{const.} : I_2 = p - \rho_0 c v = \text{const.}$$

Problem. Let at an initial instance of time $t = 0$ in an infinite pipeline $(-\infty < x < +\infty)$ there be distributions of pressure $p(x, 0) = \Pi(x)$ and fluid velocity $v(x, 0) = \Phi(x)$. It is required to determine, what motion appears in the pipeline at $t > 0$, i.e. it is required to find the functions $p(x, t)$ and $v(x, t)$ satisfying Eqs. (5.10) and (5.11) and initial conditions $p(x, 0) = \Pi(x)$, $v(x, 0) = \Phi(x)$.

Solution. Consider a plane (x, t) depicted in Fig. 5.1. Let $M(x, t)$ be an arbitrary chosen point of this plane at $t > 0$. Draw from the point $M(x, t)$ straight lines (characteristics) MA: $x - ct = x_1$ and MB: $x + ct = x_2$.

Since at the characteristic $x - ct = x_1$ the first Riemann invariant I_1 is constant, we can write

$$p_M(x, t) + \rho_0 c \cdot v_M(x, t) = p_A(x_1, 0) + \rho_0 c \cdot v_A(x_1, 0).$$

At the characteristic $x + ct = x_2$ the second Riemann invariant I_2 is constant

$$p_M(x, t) - \rho_0 c \cdot v_M(x, t) = p_A(x_2, 0) - \rho_0 c \cdot v_A(x_2, 0).$$

Resolving the system of obtained equations relative to $p_M(x, t)$ and $v_M(x, t)$, we get

$$\begin{cases} p_M(x, t) = \dfrac{p(x_1, 0) + p(x_2, 0)}{2} + \rho_0 c \cdot \dfrac{v(x_1, 0) - v(x_2, 0)}{2}, \\ v_M(x, t) = \dfrac{p(x_1, 0) - p(x_2, 0)}{2\rho_0 c} + \dfrac{v(x_1, 0) + v(x_2, 0)}{2} \end{cases}$$

or

$$\begin{cases} p_M(x, t) = \dfrac{\Pi(x_1) + \Pi(x_2)}{2} + \rho_0 c \cdot \dfrac{\Phi(x_1) - \Phi(x_2)}{2}, \\ v_M(x, t) = \dfrac{\Pi(x_1) - \Pi(x_2)}{2\rho_0 c} + \dfrac{\Phi(x_1) + \Phi(x_2)}{2} \end{cases}$$

Substituting in these relations instead of x_1 and x_2 their expression through x and t, we receive the solution of the problem

$$\begin{cases} p_M(x,t) = \dfrac{\Pi(x-ct)+\Pi(x+ct)}{2} + \rho_0 c\cdot\dfrac{\Phi(x-ct)-\Phi(x+ct)}{2}, \\ v_M(x,t) = \dfrac{\Pi(x-ct)-\Pi(x+ct)}{2\rho_0 c} + \dfrac{\Phi(x-ct)+\Phi(x+ct)}{2} \end{cases} \tag{5.14}$$

Since functions $\Pi(x)$ and $\Phi(x)$ are known, the problem is completely solved. Equalities (5.14) are called the *d'Alambert formulas.*

5.3.3
Propagation of Waves in a Semi-Infinite Pipeline

A pipeline is called *semi-infinite* when it has the initial cross-section ($x = 0$) and runs from it in the positive direction of the x-axis ($x > 0$) to infinity. It is a *model* of a pipeline in which the conditions at one of the ends (left end) are taken into account whereas the influence of another end (right end) is neglected.

Problem. Let, at the initial instant of time $t = 0$ the fluid in a semi-infinite ($0 < x < +\infty$) pipeline be quiescent $v(x, 0) = 0$ and the pressure constant $p(x, 0) = p_0$. The fluid velocity in the initial cross-section $x = 0$ at $t > 0$ suddenly begins to change with a law $v(0, t) = \Phi(t)$. It is required to determine the pattern of fluid flow in the pipeline at $t > 0$.

Solution. It is required to determine what kind of velocity and pressure waves begin to propagate in a semi-infinite pipeline caused by perturbations created at the left end of the pipeline.

Consider a plane (x, t) at $t > 0$, $x \geq 0$ as depicted in Fig. 5.2. Draw through the origin of the coordinates in this plane the characteristic $x = ct$ with positive slope. At points N of this plane located below the straight line $x = ct$ the d'Alambert formulas (5.14) give the following equations

$$\begin{cases} p_N(x,t) = \dfrac{p_0+p_0}{2} + \rho_0 c\cdot\dfrac{0-0}{2} = p_0, \\ v_N(x,t) = \dfrac{p_0-p_0}{2\rho_0 c} + \dfrac{0+0}{2} = 0 \end{cases} \tag{5.15}$$

This result has a simple physical meaning. In the region $x > ct$ of the pipeline, to which the instant of time t has not yet come, a perturbation (signal) is propagated with velocity c from the initial pipeline cross-section, the fluid is at rest as before, that is its velocity is 0 and the pressure is p_0.

Consider now the region $x < ct$ of the pipeline to which at time t have come perturbations from the initial cross-section of the pipeline. Draw through a

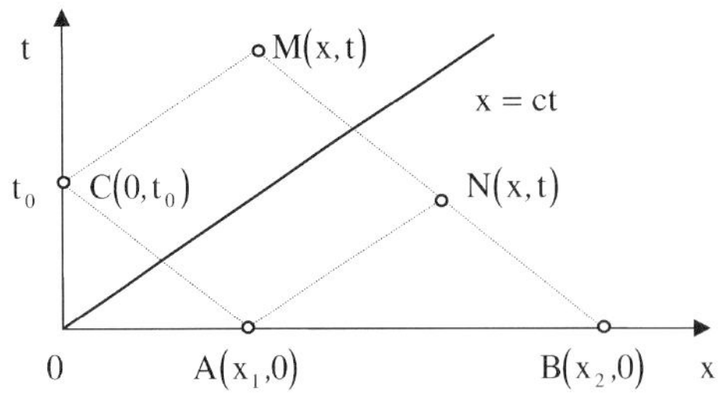

Figure 5.2 Diagram illustrating the problem on traveling waves.

point M(x, t) of this region two characteristics MC and MB and from the point C, located at initial cross-section of the pipeline (time t_0), the characteristic CA (Fig. 5.2).

With the help of the condition at the characteristic CA consisting in constancy of the second Riemann invariant, and relevant boundary condition we find the pressure in the initial pipeline cross-section at time t_0

$$p_C(0, t_0) - \rho_0 c \cdot v_C(0, t_0) = p_A(x_1, 0) - \rho_0 c \cdot v_C(x_1, 0) = p_0 - 0 = p_0.$$

From this follows

$$p_C(0, t_0) = p_0 + \rho_0 c \cdot v_C(0, t_0).$$

Hence the pressure p_C at the initial pipeline cross-section, having been initially equal to p_0, has varied over $\rho_0 c \cdot v_C(0, t_0)$ as a result of the increase in velocity $v_C(0, t_0)$ in this cross-section.

Since the pressure and velocity of fluid at points C and B are now known, one can find the pressure at the point M(x, t)

$$\begin{cases} p_M(x, t) + \rho_0 c \cdot v_M(x, t) = p_C + \rho_0 c \cdot v_C = (p_0 + \rho_0 c \cdot v_0) + \rho_0 c \cdot v_0, \\ p_M(x, t) - \rho_0 c \cdot v_M(x, t) = p_B - \rho_0 c \cdot v_B = p_0 - \rho_0 c \cdot 0 = p_0 \end{cases}$$

From this follows

$$\begin{cases} p_M(x, t) = p_0 + \rho_0 c \cdot v_C(0, t_0), \\ v_M(x, t) = v_C(0, t_0) \end{cases} \tag{5.16}$$

Note that t_0 is determined through the coordinates (x, t) of the point M by the formula $t_0 = t - x/c$ (the characteristic equation for MC is $x = c \cdot (t - t_0)$). Therefore relations (5.16) have the final form

$$\begin{cases} p_M(x, t) = p_0 + \rho_0 c \cdot v_C\left(0, t - \dfrac{x}{c}\right) = p_0 + \rho_0 c \cdot \Phi\left(t - \dfrac{x}{c}\right), \\ v_M(x, t) = \Phi\left(t - \dfrac{x}{c}\right), \end{cases} \tag{5.17}$$

where Φ is a function determining the velocity change at the initial ($x = 0$) pipeline cross-section not at the instant of time t but at a later time $t - x/c$.

It is clear that the quantity x/c is equal to the time in which the perturbation (signal) reaches the considered cross-section x from the initial one.

Formulas (5.17) show that any velocity change at the initial pipeline cross-section propagates rightwards along the pipeline as a traveling wave giving rise to a pressure traveling wave exceeding the initial value p_0 by $\rho_0 \tilde{n} \cdot v_C$. Formulas (5.15) and (5.17) give the complete solution of the considered problem.

Exercise. The pumping in quiescent diesel fuel ($\rho_0 = 840$ kg m^{-3}, $c = 1060$ m s^{-1}) in a semi-infinite pipeline has begun with constant flow rate so that $v_0 = 1.5$ m s^{-1}. It is required to determine by how much the pressure has increased at the pumping cross-section.

Solution. The first formula (5.17) gives $p(0, t) - p_0 = \rho_0 c \cdot v(0, t) = \rho_0 c v_0 = 840 \cdot 1060 \cdot 1.5 \cong 1.34 \cdot 10^6$ MPa or approximately 13.6 atm.

Answer. 1.34 MPa.

5.3.4
Propagation of Waves in a Bounded Pipeline Section

Consider the problem on wave interaction in a limited pipeline section ($0 \leq x \leq L$) between the initial cross-section $x = 0$ and the final cross-section $x = L$ assuming the absence of friction. In such a problem we need, besides the *initial conditions* $p(x, 0)$ and $v(x, 0)$, *boundary conditions* reflecting the interaction of the pipeline section under consideration with equipment located at the pipeline ends, that is at $x = 0$, $t > 0$ and $x = L$, $t > 0$.

Now we should say something about the number of these conditions. Through each point of the left pipeline section boundary $x = 0$, $t > 0$ passes only one characteristic of the wave equation, namely the characteristic of negative slope $x = -ct + \text{const}$. Along it the condition $p - \rho_0 c \cdot v = \text{const.}$ should be obeyed. Thus at points on this boundary there is always one algebraic relation between the quantities p and v. To determine a unique solution for p and we need an additional relation, that is *one more boundary condition.*

Analogously, through each point of the right boundary $x = L$, $t > 0$ of the pipeline section goes only one characteristic with positive slope $x = ct + \text{const.}$ Along this characteristic the compatibility condition $p + \rho_0 c \cdot v = \text{const.}$ is obeyed. Thus at these points there should be an additional boundary condition.

The form of the boundary conditions *could be of a great variety*, dependent on the type of equipment set at the end cross-sections of the pipeline section under consideration. For example, at the left end of the pipeline can be placed a piston pump providing constant delivery of fluid into the pipeline. Then the boundary condition at $x = 0$ would be $v(0, t) = v_0 = \text{const.}$ at all $t > 0$. If the right end of the pipeline is open, the boundary condition at the right end would be $p(0, t) = p_0 = \text{const.}$ at $t > 0$. Of course there are also many other possible boundary conditions.

Problem. Let a fluid in the pipeline section $0 \le x \le L$ be initially (at $t = 0$) quiescent ($v(x, 0) = 0$) and the pressure constant ($p(x, 0) = p_0$). At $t > 0$ the fluid starts to be delivered into the pipeline by the law $v(0, t) = \Phi(t)$. The end cross-section $x = L$ of the pipeline is open to the atmosphere, so that the pressure at the cross-section is held constant $p(L, t) = p_0$. It is required to determine the motion generated in the pipeline at $t > 0$ (Fig. 5.3).

Solution. Consider on the plane (x, t) a strip $0 \le x \le L$, $t > 0$ corresponding to the variability domain of the problem (Fig. 5.3).

1. The solution in the region 1 enclosed by the triangle OCL, i.e. the region which the perturbation has not yet reached, is as follows: $p(x, t) = p_0$; $v(x, t) = 0$.
2. Find now the solution in the region 2 restricted by the triangle OCD ($2L/c$ is the time of the wave double path lengthwise in the pipeline section).

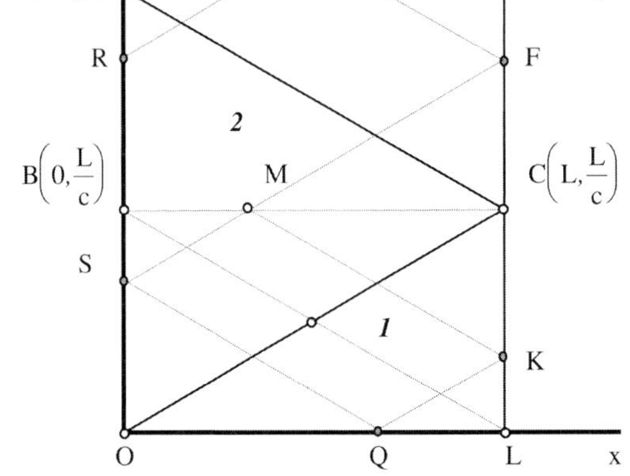

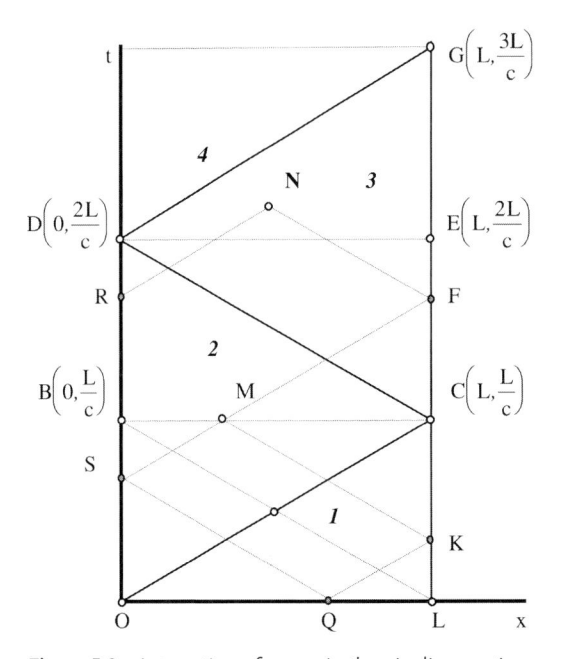

Figure 5.3 Interaction of waves in the pipeline section.

Determine first the pressure p_S at an arbitrary point S of the left boundary

$$p_S - \rho_0 c \cdot v_S = p_Q - \rho_0 c \cdot v_Q = p_0 - 0,$$

$$p_S(0, t) = p_0 + \rho_0 c \cdot v_S(t) = p_0 + \rho_0 c \cdot \Phi(t).$$

Then obtain the pressure and velocity at an arbitrary point $M(x, t)$ of the region under consideration

$$\begin{cases} p_M + \rho_0 c \cdot v_M = p_S + \rho_0 c \cdot v_S = p_0 + 2\rho_0 c \cdot \Phi(t_S), \\ p_M - \rho_0 c \cdot v_M = p_0 - \rho_0 c \cdot 0 = p_0 \end{cases}$$

from which follows

$$\begin{cases} p_M(x, t) = p_0 + \rho_0 c \cdot \Phi\left(t - \dfrac{x}{c}\right), \\ v_M(x, t) = \Phi\left(t - \dfrac{x}{c}\right) \end{cases}$$

3. Find the solution in the region 3 bounded by the triangle CDG ($3L/c$ is the time of the wave triple path lengthwise in the pipeline section).
   Determine first the fluid velocity v_F at an arbitrary point F of the right boundary

$$p_F + \rho_0 c \cdot v_F = p_S + \rho_0 c \cdot v_S = [p_0 + \rho_0 c \cdot \Phi(t_S)] + \rho_0 c \cdot \Phi(t_S)$$

Since $p_F = p_0$ and $t_S = t - L/c$, then $v_F(L, t) = 2 \cdot \Phi(t - L/c)$.
   After this we get the pressure and velocity at the arbitrary point $N(x, t)$ of the considered region

$$\begin{cases} p_N + \rho_0 c \cdot v_N = p_R + \rho_0 c \cdot v_R = p_0 + 2\rho_0 c \cdot \Phi\left(t - \dfrac{x}{c}\right), \\ p_N - \rho_0 c \cdot v_N = p_F - \rho_0 c \cdot v_F = p_0 - 2\rho_0 c \cdot \Phi\left(t - \dfrac{x}{c}\right) \end{cases}$$

Hence, $p_N = p_0$; $v_N = 2\Phi(t - x/c)$.

In a similar manner by the method of characteristics the solution in regions 4, 5 and other regions of the strip under consideration could be found.

5.3.5 Method of Characteristics

Let us return to the system of equations (5.4) describing non-stationary flow of a slightly compressible fluid with regard to viscous friction forces

$$\begin{cases} \dfrac{\partial p}{\partial t} + \rho_0 c^2 \cdot \dfrac{\partial v}{\partial x} = 0, \\ \rho_0 \dfrac{\partial v}{\partial t} + \dfrac{\partial p}{\partial x} = -\lambda(Re, \varepsilon) \dfrac{1}{d_0} \dfrac{\rho_0 v^2}{2} - \rho_0 g \sin \alpha(x). \end{cases}$$

Multiplication of the second equation by c and addition of the result to the first equation yields

$$\left(\frac{\partial p}{\partial t}+c\frac{\partial p}{\partial x}\right)+\rho_0 c\cdot\left(\frac{\partial v}{\partial t}+c\frac{\partial v}{\partial x}\right)=-\lambda\frac{\rho_0 c v|v|}{2d}-c\rho_0 g\,\sin\alpha.$$

In a similar manner after subtraction of the second equation multiplied by c from the first one we obtain

$$\left(\frac{\partial p}{\partial t}-c\frac{\partial p}{\partial x}\right)-\rho_0 c\cdot\left(\frac{\partial v}{\partial t}-c\frac{\partial v}{\partial x}\right)=\lambda\frac{\rho_0 c v|v|}{2d}+c\rho_0 g\,\sin\alpha.$$

If at the plane (x, t) we consider straight lines determined by the equations

1. $\dfrac{dx}{dt}=c, \quad x-ct=\xi=\text{const.},$
2. $\dfrac{dx}{dt}=-c, \quad x+ct=\eta=\text{const.},$

which for the wave equation are called *characteristics*, it is easy to reveal that for any parameter $A(x, t)$

$$\frac{\partial A}{\partial t}+c\cdot\frac{\partial A}{\partial x}=\left(\frac{dA}{dt}\right)_{\xi=\text{const.}}$$

is valid.

This means that the expression on the left-hand side is the derivative of the function $A(x, t)$ in the direction of the first characteristic ($\xi = \text{const.}$). Similarly it is true that

$$\frac{\partial A}{\partial t}-c\cdot\frac{\partial A}{\partial x}=\left(\frac{dA}{dt}\right)_{\eta=\text{const.}}$$

i.e. the expression on the left-hand side is the derivative of the function $A(x, t)$ in the direction of the second characteristic ($\eta = \text{const.}$).

Using now the notion of a directional derivative, one can write the above obtained equations as follows

$$\left(\frac{dp}{dt}\right)_{\xi=\text{const.}}+\rho_0 c\cdot\left(\frac{dv}{dt}\right)_{\xi=\text{const.}}=-\lambda\frac{\rho_0 c v|v|}{2d}-\rho_0 c g\,\sin\alpha,$$

$$\left(\frac{dp}{dt}\right)_{\eta=\text{const.}}-\rho_0 c\cdot\left(\frac{dv}{dt}\right)_{\eta=\text{const.}}=\lambda\frac{\rho_0 c v|v|}{2d}+\rho_0 c g\,\sin\alpha,$$

or

$$\begin{cases}\dfrac{d}{dt}(p+\rho_0 c\cdot v)_{\xi=\text{const.}}=-\lambda\dfrac{\rho_0 c v|v|}{2d}-\rho_0 c g\,\sin\alpha,\\[2ex] \dfrac{d}{dt}(p-\rho_0 c\cdot v)_{\eta=\text{const.}}=\lambda\dfrac{\rho_0 c v|v|}{2d}+\rho_0 c g\,\sin\alpha.\end{cases} \tag{5.18}$$

At $\lambda = 0$ and $\alpha = 0$ the right-hand sides of Eqs. (5.18) vanish. This means that along the characteristics of positive slope is conserved the quantity $I_1 = p_0 + \rho_0 c \cdot v$, whereas the quantity $I_2 = p_0 - \rho_0 c \cdot v$ is conserved along the characteristic of negative slope. This conclusion is consistent with the results obtained above for the wave equation.

If $\lambda \neq 0$, the quantities I_1 and I_2 are not constants at relevant characteristics. Nevertheless, Eqs. (5.18) may be used to calculate various non-stationary flows in a pipeline, especially when numerical methods are used.

Let for example at the instant of time t_{m-1} (in particular at $t = 0$), the distributions of pressure $p = p(x, t_{m-1})$ and velocity $v = v(x, t_{m-1})$ in the pipeline be known. Show how the values of these parameters at the next instant of time $t_m = t_{m-1} + \Delta t$ could be calculated.

Consider on the plane (x, t) a rectangular grid with coordinate step Δx and time step $\Delta t = \Delta x / c$ (Fig. 5.4). Through the nodes of the resulting grid let us draw characteristics $x = ct + \text{const.}$ and $x = -ct + \text{const.}$ of positive and negative slope, respectively. Continuous distribution of the sought functions $p(x, t)$ and $v(x, t)$ is replaced with discrete values $p_{k,m} = p(x_k, t_m)$ and $v_{k,m} = v(x_k, t_m)$ of the grid functions at the grid nodes. Suppose that all values $p_{k,m-1}$ and $v_{k,m-1}$ are known and it is required to find the values $p_{k,m}$ and $v_{k,m}$ of grid functions at $t = t_m$.

Let $\mathrm{M}(x_k, t_m)$ be an arbitrary point on the plane (x, t). Replacing the directional derivatives in Eqs. (5.18) with finite differences along the characteristics AM and BM we obtain

$$\begin{cases} \left.\dfrac{\Delta(p + \rho_0 c \cdot v)}{\Delta t}\right|_{\xi=\text{const.}} = -c \cdot \phi_{\mathrm{A}}, \\ \left.\dfrac{\Delta(p - \rho_0 c \cdot v)}{\Delta t}\right|_{\eta=\text{const.}} = c \cdot \phi_{\mathrm{B}}, \end{cases}$$

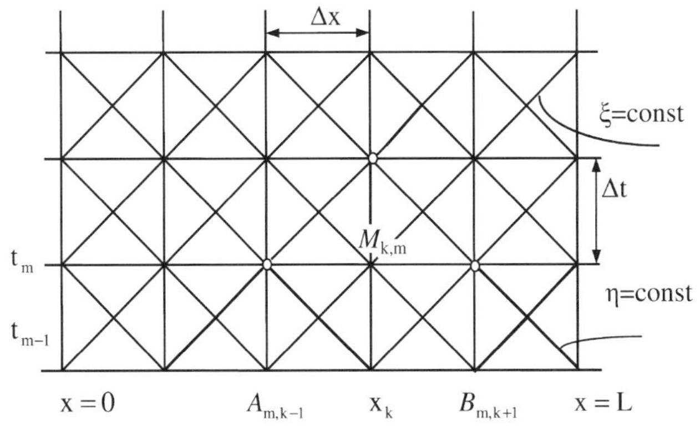

Figure 5.4 Design diagram of the characteristic method.

where $\phi = \lambda \frac{\rho_0 v|v|}{2d} + \rho_0 g \cdot \sin\,\alpha$. In addition

$$\Delta(p + \rho_0 c \cdot v)\big|_{\xi=\text{const.}} = (p_M + \rho_0 c \cdot v_M) - (p_A + \rho_0 c \cdot v_A)$$

$$\Delta(p - \rho_0 c \cdot v)\big|_{\eta=\text{const.}} = (p_M - \rho_0 c \cdot v_M) - (p_B - \rho_0 c \cdot v_B).$$

From this follows a system of equations to determine the pressure p_M and velocity v_M of the fluid at point M through known values of these parameters at points A and B

$$\begin{cases} p_M + \rho_0 c \cdot v_M = p_A + \rho_0 c \cdot v_A - \Delta t \cdot c\phi_A \\ p_M - \rho_0 c \cdot v_M = p_B - \rho_0 c \cdot v_B + \Delta t \cdot c\phi_B. \end{cases}$$

or

$$\begin{cases} p_{k,m} + \rho_0 c \cdot v_{k,m} = p_{k-1,m-1} + \rho_0 c \cdot v_{k-1,m-1} - \Delta x \cdot \phi_{k-1,m-1} \\ p_{k,m} - \rho_0 c \cdot v_{k,m} = p_{k+1,m-1} - \rho_0 c \cdot v_{k+1,m-1} + \Delta x \cdot \phi_{k+1,m-1}. \end{cases}$$

From this system of linear equations we get the pressure $p_{k,m}$ and the velocity $v_{k,m}$ of the flow

$$\begin{cases} p_{k,m} = \dfrac{p_{k-1,m-1} + p_{k+1,m-1}}{2} + \rho_0 c \cdot \dfrac{v_{k-1,m-1} - v_{k+1,m-1}}{2} \\ \qquad + \dfrac{\Delta x}{2} \cdot (\phi_{k+1,m-1} - \phi_{k-1,m-1}) \\ v_{k,m} = \dfrac{p_{k-1,m-1} - p_{k+1,m-1}}{2\rho_0 c} + \dfrac{v_{k-1,m-1} + v_{k+1,m-1}}{2} \\ \qquad - \dfrac{\Delta x}{2\rho_0 c} \cdot (\phi_{k-1,m-1} + \phi_{k+1,m-1}) \end{cases} \tag{5.19}$$

Hence, the recurrent formulas (5.19) give the solution of the formulated problem, because they allow one to calculate the pressure and velocity of the flow at the *following* instant of time t_m from known values of these parameters at the *preceding* instant of time t_{m-1}. Since at the first instant of time we can take the initial values of the pressure and velocity of the flow at $t = 0$, then calculating successively the pressure and velocity with formulas (5.19), we can get flow parameters at an arbitrary instant of time $t > 0$.

5.3.6
Initial, Boundary and Conjugation Conditions

Let us investigate non-stationary fluid flow at a pipeline section $0 \leq x \leq L$ starting from a certain instant of time $t = 0$ taken as the initial time. In order to know how the non-stationary process progresses it is necessary to have information about the initial and boundary conditions, that is to know the state of the flow before starting and what is happening at the edges of the pipeline sections, e.g. at the cross-section $x = 0$ and $x = L$. The first information is called the *initial conditions* and the second the *boundary conditions*.

Initial Conditions

The state of the pipeline section at the initial instant of time can be arbitrary, but often as the initial state is taken the stationary fluid flow existing in the pipeline at the initial instant of time.

Let, for example, for stationary flow, the known fluid flow rate be $Q = Q(x, 0) = Q_0$ and the distribution of head be $H(x, 0) = H_0 - i_0 \cdot x$, where $H_0 = H(0, 0)$ is the head at the beginning of the pipeline section; $i_0 = (H_0 - H_k)/L$, where $H_k = H(L, 0)$ is the head at the end of the pipeline section. Then as initial conditions ($m = 1$, $t_{m-1} = t_0 = 0$) may be accepted

$$v(x, 0) = \frac{Q_0}{S_0} = \text{const.};\; p(x, 0) = \rho_0 g \cdot [H_0 - i_0 \cdot x - z(x)]$$

or

$$v_{k,1} = \frac{Q_0}{S_0} = \text{const.};\; p_{k,1} = \rho_0 g(H_0 - i_0 \cdot x_k - z_k)$$

where $k = 1, 2, \ldots, N+1$, $x_k = (k-1) \cdot \Delta x$, $x_1 = 0$, $x_{N+1} = L$, $z_k = z(x_k)$, $\Delta x = L/N$ (N being the number of parts into which the pipeline section is divided).

Boundary Conditions

The formulas (5.19) permit one to find p and v at any point on the strip $0 < x < L, t > 0$ determining the pipeline section except its edges – beginning ($x = 0$) and ending ($x = L$).

Only one characteristic of the negative slope $dx/dt = -c$ comes from the integration domain to a point $M(x = 0)$ of the left pipeline section boundary (Fig. 5.5a). It gives one condition for two unknown quantities $p_{1,m}$ and $v_{1,m}$

$$p_{1,m} - \rho_0 c \cdot v_{1,m} = p_{2,m-1} - \rho_0 c \cdot v_{2,m-1} + \Delta x \cdot \phi_{2,m-1}$$

therefore an additional condition is needed. Such a condition can be an algebraic equation $F(p, v) = 0$ expressing the relation between pressure $p_M(0, t)$ and velocity $v_M(0, t)$ at the initial cross-section of the pipeline. As a rule this condition models the operation of a pumping station and is represented by its $(Q - \Delta H)$ characteristic. Thus, boundary conditions at points on the left pipeline section boundary may be represented as a system of equations

$$x = 0, t > 0: \begin{cases} p_{1,m} - \rho_0 c \cdot v_{1,m} = p_{2,m-1} - \rho_0 c \cdot v_{2,m-1} + \Delta x \cdot \varphi_{2,m-1} \\ F(p_{1,m}, v_{1,m}) = 0. \end{cases}$$

Similarly, only one characteristic of the positive slope $dx/dt = +c$ comes from the domain of integration to a point $E(x = L)$ of the right pipeline section boundary (Fig. 5.5b), Therefore the boundary condition at points on

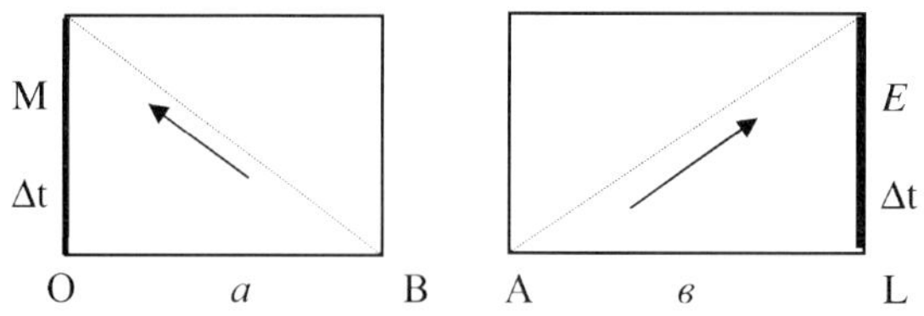

Figure 5.5 Calculation of p and v at boundary cross-sections.

the right pipeline section boundary may be represented as a system of two equations:

$$x = L, t > 0:$$
$$\begin{cases} p_{N+1,m} + \rho_0 c \cdot v_{N+1,m} = p_{N,m-1} + \rho_0 c \cdot v_{N,m-1} - \Delta x \cdot \varphi_{N,m-1} \\ G(p_{N+1,m}, v_{N+1,m}) = 0. \end{cases}$$

The dependence $G(p, v) = 0$ expresses the relation between the pressure $p_N(L, t)$ and velocity $v_N(L, t)$ at the end of the pipeline section. It should be noted that there are also possibilities for more complicated boundary conditions.

Conjugation Conditions

If the equipment responsible for process non-stationarity is located inside the pipeline section, e.g. at the cross-section x_*, $0 < x_* < L$, then at this cross-section there can exist a discontinuity of hydraulic parameters. Such a discontinuity requires additional conditions called *conjugation conditions.*

Let take place, for example, at a pipeline cross-section x_* ejection or injection of fluid with flow rate q ($q < 0$ ejection; $q > 0$ injection). Then such a cross-section is characterized by continuity of pressure and discontinuity of flow rate or velocity. Let us denote the values of the parameters before ejection or injection with superscript (−) and after ejection or injection with superscript (+). Then at the cross-section x_* the following conditions should be obeyed:

$$p^+(x_*, t) = p^-(x_*, t);\ v^+(x_*, t) - v^-(x_*, t) = \frac{q}{S}.$$

Then to calculate the three unknown parameters $p_{k,m}$, $v^+_{k,m}$, $v^-_{k,m}$ of non-stationary flow at point $x_* = x_k$ we use the following system of three linear equations

$$\begin{cases} p_{k,m} + \rho_0 c \cdot v^-{}_{k,m} = p_{k-1,m-1} + \rho_0 c \cdot v_{k-1,m-1} - \Delta x \cdot \phi_{k-1,m-1} \\ p_{k,m} - \rho_0 c \cdot v^+{}_{k,m} = p_{k+1,m-1} - \rho_0 c \cdot v_{k+1,m-1} + \Delta x \cdot \phi_{k+1,m-1} \\ v^+{}_{k,m} - v^-{}_{k,m} = \dfrac{q}{S_0} \end{cases}$$

with $p^+{}_{k,m} = p^-{}_{k,m} = p_{k,m}$.

In the case when a gate valve is located at the cross-section $x_* = x_k$, its operation is modeled by conjugation conditions

$$\begin{cases} v^+(x_*, t) = v^-(x_*, t) \\ p^-(x_*, t) - p^+(x_*, t) = \varsigma(t) \cdot \dfrac{\rho_0 v^2(x_*, t)}{2} \end{cases}$$

where $\varsigma(t)$ is the local resistance factor that varies during valve closing or opening. The first condition means the continuity of the flow rate, whereas the second condition signifies pressure discontinuity at different sides of the valve. Thus, the model of the valve is represented by the following system of three equations

$$\begin{cases} p^-{}_{k,m} + \rho_0 c \cdot v_{k,m} = p_{k-1,m-1} + \rho_0 c \cdot v_{k-1,m-1} - \Delta x \cdot \phi_{k-1,m-1} \\ p^+{}_{k,m} - \rho_0 c \cdot v_{k,m} = p_{k+1,m-1} - \rho_0 c \cdot v_{k+1,m-1} + \Delta x \cdot \phi_{k+1,m-1} \\ p^+{}_{k,m} - p^-{}_{k,m} = \varsigma(t_m) \cdot \rho_0 v^2{}_{k,m}/2 \end{cases}$$

for the three parameters $p^+{}_{k,m}$, $p^-{}_{k,m}$, $v_{k,m}$ with $v^+{}_{k,m} = v^-{}_{k,m} = v_{k,m}$.

5.3.7 Hydraulic Shock in Pipes

In all the above stated it was assumed by default that the functions $\rho(x, t)$, $p(x, t)$ and $v(x, t)$ in the differential equations are differentiable with respect to time and coordinate and in any case are certainly continuous. Nevertheless, in engineering there are processes in which these functions vary very quickly with time and in space. An example of such a process is *hydraulic shock*. The essence of hydraulic shock is that the stationary flow of fluid in a pipeline is disturbed by the abrupt closing or opening of a gate valve, the switching on or switching off of a pump and so on, resulting in hard braking or acceleration of the fluid and shock compression of the fluid particles. The front at which the variation of the hydrodynamic parameters of the fluid takes place has a relatively small extent and propagates in the form of a pressure wave down-stream and up-stream of the fluid.

Similar phenomena occur in the pipeline in other cases when the velocity (flow rate) of the fluid varies in a stepwise manner. The possibility of hydraulic shock should be taken into account in the exploitation of pipelines, since shock pressure can far exceed permissible standards, leading to pipe breakage and an emergency situation.

The explanation of hydraulic shock was given by Joukovski in his article "On hydraulic shock in water-supply pipes" (1899). He connected the magnitude of the pressure jump $[p]$ with the properties of fluid compressibility and the elasticity of the pipe and obtained the following formula

$$[p] = \rho_0 D \cdot [v] \tag{5.20}$$

where D is the velocity of shock wave propagation in the pipeline (see Eq. (5.2)) and $[v]$ the magnitude of the stepwise change in the fluid.

It should be noted that the introduction of stepwise variations (jumps) of hydrodynamic flow parameters is nothing more than a *model* of the phenomenon under consideration. In fact each such discontinuity has a transition region, though very narrow, from the value of parameter A^+ to the left of the discontinuity front up to the value A^- of the same parameter to the right of the front. The quantity $[A] = A^+ - A^-$ is called the *jump* of parameter A at the discontinuity front. To describe the structure of this transition zone needs as a rule a more complicated model than the given one. Nevertheless, introduction of the system of equations (5.4) into consideration of discontinuity solutions has proved to be very fruitful.

Conditions at Discontinuities (Jumps) of Hydrodynamic Parameters

Now let us obtain conditions which should satisfy the flow parameters at the discontinuity front. Let the discontinuity front of the flow parameters propagate with velocity $D = dx_{\hat{o}}/dt$ in the positive direction of the x-axis (Fig. 5.6).

The wave of the hydraulic shock is characterized by the parameters of flow (and pipeline cross-section area) that are subjected to discontinuities, or *jumps*, at their mobile front. However the limiting values of these parameters before ρ^-, v^-, p^-, S^- and after ρ^+, v^+, p^+, S^+ the wave front cannot be prescribed arbitrarily. They have to obey the conditions of fluid mass and momentum conservation.

In the time interval Δt all fluid particles present at the beginning of this interval at a distance $(D - v^-) \cdot \Delta t$ before the front pass through the front. Therefore the mass of fluid flowing in the wave in time Δt is determined by the expression $\rho^- S^- (D - v^-) \cdot \Delta t$.

In the same time interval a fluid mass equal to $\rho^+ S^+ (D - v^+) \cdot \Delta t$ flows out from the wave.

It is evident that the mass of fluid flowing into and out of the wave front should be equal

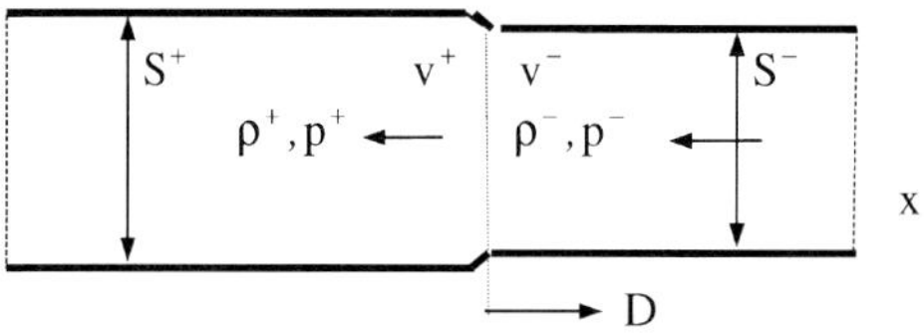

Figure 5.6 Hydraulic shock in a pipeline.

$$\rho^- S^-(D - v^-) = \rho^+ S^+(D - v^+)$$

or

$$\rho^+ v^+ S^+ - D\rho^+ S^+ = \rho^- v^- S^- - D\rho^- S.$$

If we denote a jump of any parameter by [ ] then the last relation maybe rewritten as

$$[\rho v S - \rho D \cdot S] = 0. \tag{5.21}$$

Now we can use the theorem that the momentum variation of the fluid mass that has passed through the wave front is equal to the impulse of the pressure force

$$\underbrace{\rho^+ S^+(D - v^+)\Delta t}_{\Delta m} \cdot v^+ - \underbrace{\rho^- S^-(D - v^-)\Delta t}_{\Delta m} \cdot v^- = (p^- - p^+)S^- \cdot \Delta t.$$

In the right-hand side of this equation the projection of the reaction force of the pipeline lateral surface onto the x-axis is taken into account.

We can write the last equation in the following form

$$(\rho^+ S^+ v^{+2} - \rho^+ S^+ v^+ D) - (\rho^- S^- v^{-2} - \rho^- S^- v^- D) = -S^- \Delta p$$

or

$$[v \cdot (\rho v S - D\rho \cdot S)] = -S^- \cdot [p].$$

Then, in accordance with equality (5.21), $[\rho v S - \rho D \cdot S] = 0$ the obtained equation may be simplified to

$$(\rho^- v^- S^- - D\rho^- S^-) \cdot [v] = -S^- \cdot [p]$$

or

$$\left(\frac{v^-}{D} - 1\right) \cdot \rho^- D \cdot [v] = -[p].$$

With regard to $\rho^- \approx \rho_0$ and the smallness of the ratio v^-/D we get the Joukovski formula

$$[p] = \rho_0 D \cdot [v]. \tag{5.22}$$

We can show that the velocity of the hydraulic shock wave D for a slightly compressible fluid coincides with the velocity of the perturbation propagation in a pipeline with elastic walls. In order to do this we can simplify the relation (5.21) as follows:

$$\rho_0 S_0[v] - D S_0[\rho] - D\rho_0[S] = 0$$

where it is assumed that owing to the smallness of the variations in ρ and S variations the following approximations can be made

$$[\rho v S] \approx \rho_0 S_0 [v]; \quad [\rho S] \approx \rho_0 [S] + S_0 [\rho].$$

The use of Eqs. (5.22), (2.22) and (2.23)

$$[v] = -\frac{1}{\rho_0 D} \cdot [p]; \quad [\rho] = \frac{\rho_0}{K} \cdot [p]; \quad [S] = \frac{\pi \cdot {d_0}^3}{4\delta \cdot E} \cdot [p],$$

yields

$$\frac{S_0}{D} \cdot [p] - D S_0 \frac{\rho_0}{K} \cdot [p] - \rho_0 D \frac{\pi \cdot {d_0}^3}{4\delta \cdot E} \cdot [p] = 0.$$

Since $[p] \neq 0$, we have

$$1 = D^2 \cdot \left(\frac{\rho_0}{K} + \frac{\rho_0 d_0}{\delta \cdot E} \right).$$

From here it follows that the absolute value of the hydraulic shock wave velocity $|D|$ is equal to the velocity of the perturbation propagation in the pipeline (Eq. (5.1))

$$|D| = \frac{1}{\sqrt{\dfrac{\rho_0}{K} + \dfrac{\rho_0 d_0}{\delta \cdot E}}} = c. \tag{5.23}$$

Exercise 1. Oil ($\rho_0 = 870$ kg m^{-3}, $K = 1.5 \cdot 10^9$ Pa) flows with velocity $v = 1.0$ m s^{-1} in a steel pipeline ($d_0 = 0.8$ m, $\delta = 10$ mm, $E = 2 \cdot 10^{11}$ Pa). It is required to determine how much the pressure rises at the pipeline cross-section before the gate valve suddenly closes.

Solution. Calculate first the velocity D of the hydraulic shock wave

$$D = c = \frac{1}{\sqrt{\dfrac{870}{1.5 \cdot 10^9} + \dfrac{870 \cdot 0.8}{0.01 \cdot 2 \cdot 10^{11}}}} \cong 1038 \text{ m s}^{-1}.$$

Then determine with Eq. (5.22) the pressure jump $[p] = \rho_0 D \cdot [v] = 870 \cdot 1038 \cdot 1.0 = 903\,060$ Pa or approximately 9.21 atm.

Answer. 0.903 MPa.

Exercise 2. It is required to determine how much the pressure rises at the initial cross-section of a steel pipeline ($d_0 = 0.516$ m, $\delta = 8$ mm, $E = 2 \cdot 10^{11}$ Pa) on abrupt switching on of the pumps providing the benzene feed ($\rho_0 = 750$ kg m^{-3}, $K = 1.3 \cdot 10^9$ Pa) with velocity $v = 1.5$ m s^{-1}.

Solution. Calculate first the velocity D of the hydraulic shock wave generated by the benzene pumping into the pipeline

$$D = c = \frac{1}{\sqrt{\dfrac{750}{1.3 \cdot 10^9} + \dfrac{750 \cdot 0.516}{0.008 \cdot 2 \cdot 10^{11}}}} \cong 1105 \text{ m s}^{-1}.$$

Then determine the pressure jump with Eq. (5.22) $[p] = \rho_0 D \cdot [v] = 750 \cdot 1105 \cdot 1.5 = 1\,243\,125$ Pa or approximately 12.7 atm.

Answer. 1.243 MPa.

Remark on carrying out calculations. Notice one interesting circumstance following from Eqs. (5.22) and (5.23). Let, for example, $D > 0$, i.e. the wave of discontinuity propagates in the positive direction of the x-axis. Then the characteristic of negative slope $x + ct =$ const. intersects the discontinuity at point I (Fig. 5.7).

Ignoring friction, we have

1. $p_{\Pi^-} - \rho_0 c \cdot v_{\Pi^-} = p_B - \rho_0 c \cdot v_B$ – the condition at the linear segment BΠ of the characteristic BM;
2. $p_{\Pi^+} - p_{\Pi^-} = \rho_0 c \cdot (v_{\Pi^+} - v_{\Pi^-})$ – the condition (5.22) at point I of the discontinuity;
3. $p_M - \rho_0 c \cdot v_M = p_{\Pi^+} - \rho_0 c \cdot v_{\Pi^+}$ – the condition at the linear segment Π^+M of the characteristic BM.

Subtraction of the first relation from the last one with regard to the second relation yields

$$p_M - \rho_0 c \cdot v_M = p_B - \rho_0 c \cdot v_B.$$

Hence the Riemann invariants (in the considered case $I_2 = p - \rho_0 c \cdot v$) are conserved at the relevant characteristics, even when these characteristics intersect the shock front.

Problem 1. *On the phases of direct hydraulic shock.* It is required to study the non-stationary flow generated in a pipeline section $0 \le x \le L$ by abrupt closing of the gate valve at the right-hand end ($x = L$) of the pipeline (*direct hydraulic shock*). It is assumed that the fluid before closing the gate ($t = 0$) was flowing

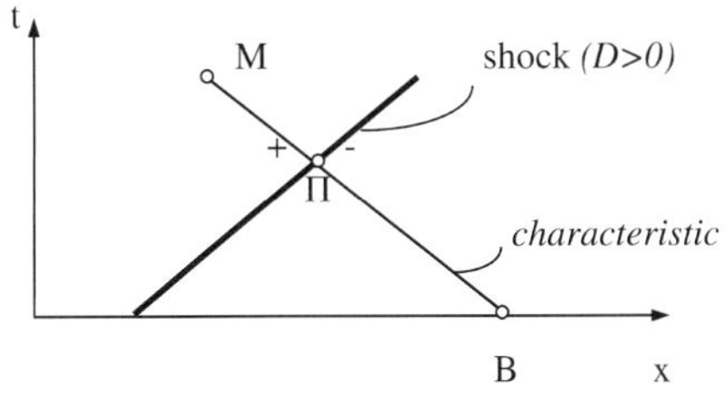

Figure 5.7 Interaction of a discontinuity with a small perturbation.

with constant velocity $v = v_0$ and friction was absent so that the pressure at all pipeline cross-sections was constant $p = p_0$. It is assumed also that at the initial cross-section ($x = 0$) the pressure is constant and equal to p_0.

Solution. The solution is shown in Fig. 5.8. By convention the solution may be divided into four phases.

1. *The phase of direct shock* continues for a time $t = L/c$. When the fluid flow is stopped, the hydraulic shock wave appears and travels from the end of the pipeline section to its beginning. The wave front brings the fluid to an abrupt stop and the pressure after the jump is raised by $\rho_0 c \cdot v_0$.
2. *The phase of the reflected wave* runs from $t_1 = L/c$ to $t_2 = 2L/c$ equal to double the time taken by the wave path along the pipeline section. In this phase the shock wave reflects at the initial cross-section and begins to move in the opposite direction. The fluid then flows out of the pipeline ($v = -v_0$) and the pressure falls to its initial value p_0.
3. *In the third phase,* lasting from the instant of time $t_2 = 2L/c$ to $t_3 = 3L/c$, the fluid continues to flow out of the pipeline with velocity $v = -v_0$ whereas the pressure in the wave reflected from the cross-section $x = L$ is lowered stepwise by $\rho_0 c \cdot v_0$ becoming lower than its original value. If the pressure is lowered to the saturated vapor

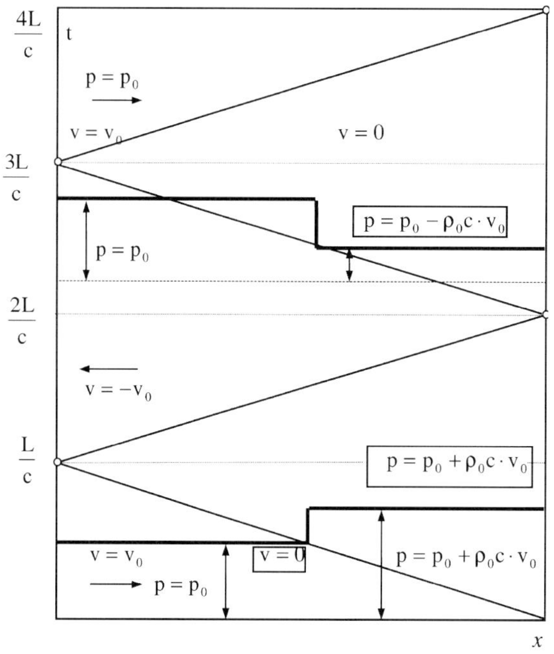

Figure 5.8 Phases of direct hydraulic shock.

pressure of the fluid, the latter would begin to boil, otherwise at $(p_0 - \rho_0 c \times v_0) > p_s$ the fluid does not vaporize and the pipeline section remains completely filled with fluid. The velocity behind the shock front propagating in the negative direction vanishes.

4. *In the fourth phase* of the process at $3L/c < t < 4L/c$ the fluid again begins to flow into the pipeline ($v = v_0$) and the pressure becomes equal to $p = p_0$. At the instant of time $t_4 = 4L/c$ the situation returns to the initial one, after which the process is *periodically* repeated.

Problem 2. *On the damping of the pressure bow shock at the hydraulic shock wave front* (Charniy, 1975). In the presence of friction the hydraulic shock wave is gradually damped in the pipeline, in particular the magnitude of the pressure jump at the shock front decreases. It is required to determine the intensity of such a fall.

Solution. Let the front of the hydraulic shock travel from the end of the pipeline cross-section $x = L$ to its beginning. Consider two negative slope characteristics parallel to the discontinuity along its left- and right-hand sides. Then in accordance with Eq. (5.18) we have

$$\frac{\mathrm{d}}{\mathrm{d}t}(p^+ - \rho_0 c \cdot v^+)\bigg|_{\dot{x}=-c} = \lambda^+ \cdot \frac{1}{d_0}\frac{\rho_0 c \cdot v^+|v^+|}{2} + \rho_0 c g \sin\alpha,$$

$$\frac{\mathrm{d}}{\mathrm{d}t}(p^- - \rho_0 c \cdot v^-)\bigg|_{\dot{x}=-c} = \lambda^- \cdot \frac{1}{d_0}\frac{\rho_0 c \cdot v^-|v^-|}{2} + \rho_0 c g \sin\alpha,$$

$$p^+ - p^- = -\rho_0 c \cdot (v^+ - v^-).$$

Subtracting term-by-term from the second equation from the first one and taking into account the third equation, we get

$$\frac{\mathrm{d}}{\mathrm{d}t}(p^+ - p^-)\bigg|_{\dot{x}=-c} = \frac{1}{2} \cdot \frac{1}{d_0}\frac{\rho_0 c}{2} \cdot \left[\lambda^- \cdot v^-|v^-| - \lambda^+ \cdot v^+|v^+|\right]$$

Taking $v^+ = v_0; v^- = v^+ - (p^+ - p^-)/\rho_0 c = v_0 - [p]/\rho_0 c$; and $[p] = p^+ - p^-$, one obtains the ordinary differential equation for the variation of the pressure jump $[p]$ at the wave front

$$\frac{\mathrm{d}[p]}{\mathrm{d}t}\bigg|_{\dot{x}=-c} = -\frac{\rho_0 c}{4 d_0} \cdot (\lambda_0 v_0{}^2 - \lambda^- v^{-2}) \tag{5.24}$$

This equation should be integrated under the initial condition $[p] = -\rho_0 c \cdot v_0$ at $t = 0$. If $\lambda^-(Re^-, \varepsilon) = \lambda^-(v^-)$ is known, the solution of the problem can be obtained without difficulty.

At $\lambda_0 v_0 \approx \lambda^- v^- = 2a > 0$, where a is a constant, the solution has a particularly simple form (for laminar flow this result gives an exact solution). We have

$$\frac{\mathrm{d}[p]}{\mathrm{d}t}\bigg|_{\dot{x}=-c} = -\frac{a \rho_0 c}{2 d_0} \cdot (v_0 - v^-) = -\frac{a}{2 d_0} \cdot [p] \tag{5.25}$$

and

$$[p] = -\rho_0 c v_0 \cdot \exp(-a/2d_0 \cdot t), \tag{5.26}$$

that is, the pressure jump at the hydraulic shock wave front decays exponentially.

From the obtained formula follows, in particular, that in pipelines of large diameter the value of the bow pressure jump at the hydraulic shock wave front decays more slowly than in pipelines of smaller diameter.

Problem 3. At the junction of two pipes with different internal diameters d_1 and d_2 comes, as viewed from the first pipe, the pressure shock wave (incident wave) with pressure amplitude of magnitude $[p_{\text{inc}}]$. There are then generated two pressure waves: one is the reflected wave, which travels in the opposite direction with amplitude $[p_{\text{refl}}]$; the other is the transmitted wave with amplitude $[p_{\text{trans}}]$, which travels through the second pipe. It is required to express the amplitudes of the reflected and transmitted waves through the amplitude of the incident wave, if the velocities of wave propagation in the first and second pipes are c_1 and c_2 respectively.

Answer. $[p_{\text{refl}}] = [p_{\text{inc}}] \cdot \dfrac{c_1 d_2^2 - c_2 d_1^2}{c_1 d_2^2 + c_2 d_1^2}$; $[p_{\text{trans}}] = [p_{\text{inc}}] \cdot \dfrac{2 c_2 d_1^2}{c_1 d_2^2 + c_2 d_1^2}$.

Remark. From the solution of the problem it follows that at $c_1 d_2{}^2 = c_2 d_1^2$ or $d_2 = d_1 \cdot \sqrt{c_2/c_1}$, $[p_{\text{ref}}] = 0$, that is, the reflected wave is not generated at the junction of the pipes.

Problem 4. It is required to prove that the amplitude of the hydraulic shock wave falling at the closed pipeline end will be doubled when reflecting from the pipeline end.

Hint. Use the results of the previous problem.

5.3.8 Accounting for Virtual Mass

In Section 5.1 it was indicated that in non-stationary processes the inertial properties of fluid in a pipeline are characterized by variable density, hence an additional quantity called *virtual mass* should be added. The main equations of non-stationary flow of a slightly compressible fluid in a pipeline with regard to virtual mass take the form

$$\begin{cases} \dfrac{\partial p}{\partial t} + \rho_0 c^2 \cdot \dfrac{\partial v}{\partial x} = 0 \\ \alpha_k \rho_0 \dfrac{dv}{dt} = -\dfrac{\partial p}{\partial x} - \lambda(Re, \varepsilon) \cdot \dfrac{1}{d} \dfrac{\rho_0 v|v|}{2} - \rho_0\, g \sin\alpha(x) \end{cases} \tag{5.27}$$

If we denote $\rho_* = \alpha_k \rho_0$ and $-c_* = -c/\sqrt{\alpha_k}$, the system of equations (5.27) transforms into

$$\begin{cases} \dfrac{\partial p}{\partial t} + \rho_* {c_*}^2 \cdot \dfrac{\partial v}{\partial x} = 0 \\ \rho_* \dfrac{dv}{dt} = -\dfrac{\partial p}{\partial x} - \lambda(Re, \varepsilon) \cdot \dfrac{1}{d} \dfrac{\rho_0 v|v|}{2} - \rho_0 g \sin \alpha(x) \end{cases} \tag{5.28}$$

equivalent to the system of equations (5.4). From this follows that the Joukowski formula (5.20) for the amplitude $[p]$ of the hydraulic shock varies; it will be enhanced by the factor $\sqrt{\alpha_\kappa}$

$$[p] = \rho_* c_* \cdot [v] = \alpha_k \rho_0 \cdot \frac{c}{\sqrt{\alpha_k}} \cdot [v] = \rho_0 c \sqrt{\alpha_\kappa} \cdot [v] \tag{5.29}$$

The velocity D of hydraulic shock wave propagation will also vary, it will be reduced by a factor $\sqrt{\alpha_\kappa}$

$$|D| = \frac{1}{\sqrt{\dfrac{\rho_*}{K} + \dfrac{\rho_* d_0}{\delta \cdot E}}} = c_* = \frac{c}{\sqrt{\alpha_\kappa}} \tag{5.30}$$

If we take into account that for turbulent flow $\alpha_k \approx 1.03$, corrections to Joukovski formula will be small (one of them (5.29) was first obtained by Leibenson *et al.*, 1934). But for laminar flow $\alpha_k \approx 1.33$ and these corrections could be significant.

5.3.9
Hydraulic Shock in an Industrial Pipeline Caused by Instantaneous Closing of the Gate Valve

In Fig. 5.9 is shown the development of the hydraulic shock in a pipeline section with diameter $D = 1020$ mm ($\delta = 10$ mm) and length $L = 100$ km upon instantaneous closing of the gate valve at the right-hand edge of the pipeline section. The pipeline is transporting crude oil with density $\rho_0 = 880$ kg m^{-3} and viscosity $\nu = 20$ cSt. At the initial instant of time $t = 0$ the flow in the pipeline is stationary with velocity $v(x, 0) = v_0 \approx 1.5$ m s^{-1}; the velocity of the hydraulic shock wave propagation c is equal to 870 m s^{-1}. The abrupt closing of the gate valve at the right-hand edge of the pipeline is modeled by the condition $v(L, t) = 0$ at $t > 0$.

Figures 5.9a and 5.9b demonstrate the distributions of the head $(z + p/\rho g)$ at 2 and 40 s, respectively, after closing the gate valve.

The initial value of the pressure jump Δp_f at the wave front is related to the initial velocity v_0 by

$\Delta p_f = \rho_0 c v_0 = 880 \cdot 870 \cdot 1.5 \cong 1.15$ MPa or ≈ 11.7 atm.

Figures 5.10a and 5.10b show the following stages of pressure wave upstream propagation.

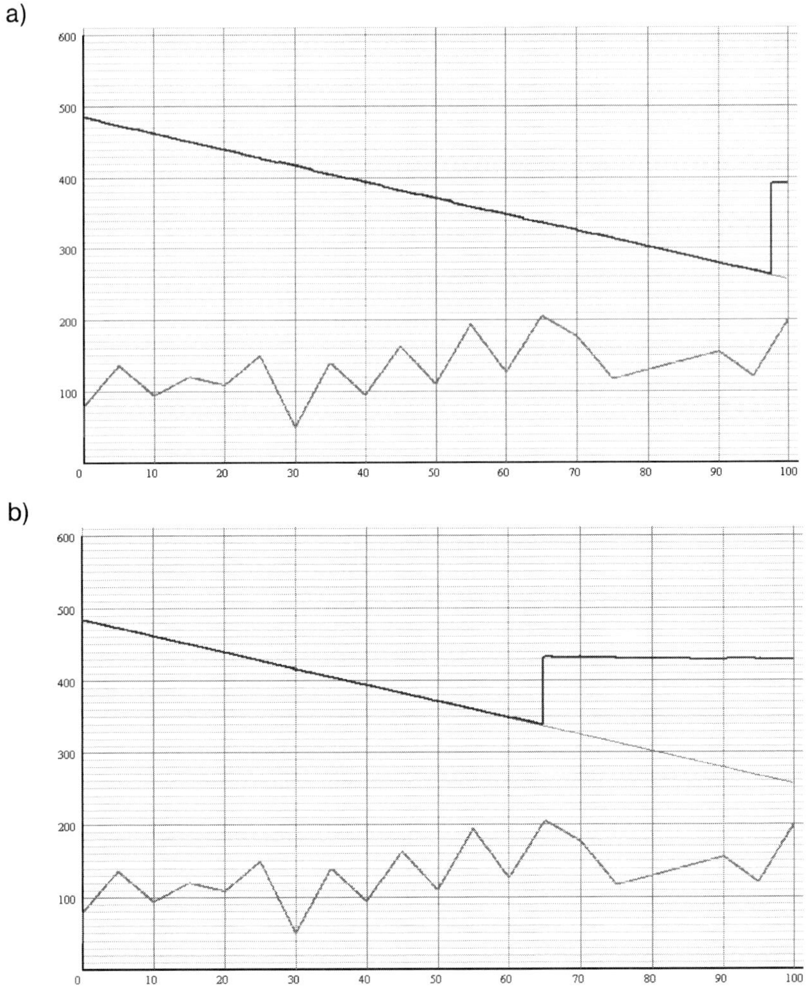

Figure 5.9 Pressure wave propagation: (a) 2 s and (b) 40 s after appearance of the wave.

The velocity $v(x, t)$ of the fluid flow behind the pressure wave front is small, therefore head losses in this region are also small. That is why the head, and consequently the pressure before the closed gate valve, are always raised. The head $H_L(t) = z_L + p_L(t)/\rho g$ at the end of the pipeline section, i.e. before the gate valve, can be taken as approximately equal to the head after the pressure wave front, that is to

$$H_L(t) = z_L + \frac{p_L(t)}{\rho_0 g} \gg H_f + \frac{[p]}{\rho_0 g},$$

where z_L is the elevation of the section end and H_f the head at the pipeline

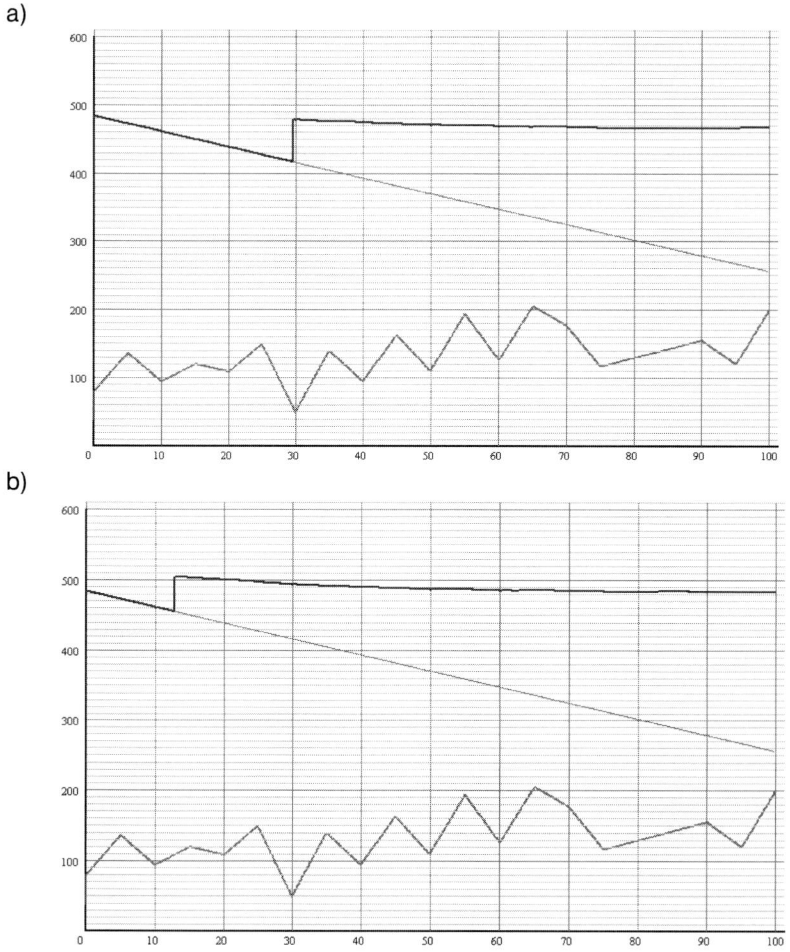

Figure 5.10 Pressure wave propagations: (a) 80 s and (b) 100 s after appearance of the wave.

cross-section which the hydraulic wave shock has reached. From this it follows that for an approximate estimation of the pressure at the pipeline section end one can use the formula

$$p_L(t) \gg \rho_0 g(H_f - z_L) + [p]$$

or

$$p_L(t) \approx \rho_0 g i_0 \cdot ct + [p], \tag{5.31}$$

where i_0 is the hydraulic gradient of the flow in the undisturbed region and t is the time elapsed after closure of the gate valve.

It should be noted that owing to viscous friction the fluid behind the wave front would not come to a halt immediately but gradually, therefore the pressure shock amplitude $[p]$ reduces monotonically (see formula (5.26)).

Protection of Pipelines from Hydraulic Shocks

The necessity to prevent the destructive force of hydraulic shock in pipelines transporting heavy dropping liquids (oil, oil products, water and other) is manifested by the fact that such pipelines, as distinct from gas-pipelines, are not equipped with cocks that close the pipeline cross-section too rapidly, but are equipped with valves, gates and slowly closing cocks. All of these should ensure safe fluid braking in the pipeline.

Pumping stations are sometimes equipped with special devices intended to protect pipelines from hydraulic shock waves. In the suction lines of pumping stations are, for example, *flow dampers* of hydraulic shock – *special safety valves* or *systems of pressure wave smoothing* in case of a sudden pumping station switch-off when the pressure before the station begins to build up. These devices operate on the principle of emergency discharge of part of the fluid from the pipeline into a special reservoir to decrease the magnitude of the pressure and its rate of increase. The safety valves open the fluid discharge when the pressure exceeds a certain value, the systems of pressure wave smoothing are switched on when the rate of pressure build-up in the suction line of the pumping station is greater than the permissible magnitude.

5.4 Non-Isothermal Gas Flow in Gas-Pipelines

Consider now non-stationary and non-isothermal gas flow in gas-pipelines. The main distinction of such flows from the flows considered above is that the gas *represents a significantly compressible medium* with density dependent not only on pressure but also on temperature. Thus, to describe these flows it is necessary not only to use the laws of mass and momentum conservation but also the law of energy transformation. In other words, as well as continuity and momentum equations, the equation of heat inflow should be invoked.

Basic equations for the calculation of non-stationary non-isothermal flows in a gas-pipeline are represented by the system (5.8) giving

$$\begin{cases} \dfrac{\partial \rho}{\partial t} + \dfrac{\partial \rho v}{\partial x} = 0 \\ \dfrac{\partial \rho v}{\partial t} + \dfrac{\partial}{\partial x}(p + \rho v^2) = -\lambda(Re, \varepsilon)\dfrac{1}{d_0}\dfrac{\rho v \cdot |v|}{2} - \rho g \sin\alpha \\ \rho\left(\dfrac{\partial e_{\text{in}}}{\partial t} + v\dfrac{\partial e_{\text{in}}}{\partial x}\right) = -\dfrac{4\kappa}{d_0}(T - T_{\text{ex}}) - p\dfrac{\partial v}{\partial x} + \lambda(Re, \varepsilon)\dfrac{1}{d_0}\dfrac{\rho v^3}{2} \end{cases} \tag{5.32}$$

in which $\rho = p/ZRT$ is the equation of the gas state and $\lambda(Re, \varepsilon)$ is the hydraulic resistance factor.

The system of equations (5.32) represents partial differential equations for three unknown functions $p(x, t)$, $v(x, t)$ and $T(x, t)$ dependent on x and t.

These non-linear equations have a complicated structure, therefore the question arises: how to solve them? To answer this question it is necessary to examine the structure of the system (5.32) treating it as well as was done when investigating equations (5.4) in the model of non-stationary flow of a slightly compressible fluid. In the course of examining this model we have seen that on the plane of variables (x, t) there are certain lines (characteristics) along which the system of partial differential equations (5.4) transforms into an ordinary differential equation providing a relation between unknown functions. Through each point $M(x, t)$ of the plane (x, t) go just two (by the number of equations) such lines, or strictly speaking

$$\frac{\mathrm{d}}{\mathrm{d}t}(p + \rho_0 c v) = I_1(p, v) \quad \text{along the line } \dot{x} = c \text{ (or } x = ct + \text{const.)},$$

$$\frac{\mathrm{d}}{\mathrm{d}t}(p - \rho_0 c v) = I_2(p, v) \quad \text{along the line } \dot{x} = c \text{ (or } x = -ct + \text{const.)}.$$

The presence of two characteristics is the distinctive property of the considered system of equations and allows us to assign the system to a class of *hyperbolic differential equations* and give a constructive method of its solution.

Let us show that the system of equations (5.32) is hyperbolic (the number of its characteristics is equal to the number of equations, that is three). We look for characteristics on the plane (x, t), i.e. lines $x = x(t)$ such that along them Eqs. (5.32) give certain differential relations between unknown functions.

Let $x = x(t)$ be a line on the plane (x, t) on which the values of the functions $p(x, t)$, $v(x, t)$, $T(x, t)$ are known. Then it is possible to write three equations relating the derivatives $\frac{\partial p}{\partial t}, \frac{\partial p}{\partial x}, \frac{\partial v}{\partial t}, \frac{\partial v}{\partial x}$ and $\frac{\partial T}{\partial t}, \frac{\partial T}{\partial x}$ of these functions along the line $x = x(t)$. To do this let us differentiate p, v, T along $x = x(t)$

$$\begin{aligned} \left.\frac{\mathrm{d}p}{\mathrm{d}t}\right|_{x(t)} &= \frac{\partial p}{\partial t} + \frac{\partial p}{\partial x} \cdot \frac{\mathrm{d}x}{\mathrm{d}t} \\ \left.\frac{\mathrm{d}v}{\mathrm{d}t}\right|_{x(t)} &= \frac{\partial v}{\partial t} + \frac{\partial v}{\partial x} \cdot \frac{\mathrm{d}x}{\mathrm{d}t} \qquad (5.33) \\ \left.\frac{\mathrm{d}T}{\mathrm{d}t}\right|_{x(t)} &= \frac{\partial T}{\partial t} + \frac{\partial T}{\partial x} \cdot \frac{\mathrm{d}x}{\mathrm{d}t} \end{aligned}$$

where $\mathrm{d}x/\mathrm{d}t$ is the slope of this line (we shall call it a characteristic) to the t-axis. The slope can be taken as given since the function $x = x(t)$ is known.

Since the left-hand sides of Eqs. (5.33) are known as well as the slope $\mathrm{d}x/\mathrm{d}t$ of the line $x(t)$, these equations could be considered as three linear equations to determine six partial derivatives p, v, T with respect to time and coordinate. If to these equations we add three equations of the system (5.32), which reduces also to linear equations with respect to the same derivatives, we get six linear equations for six partial derivatives

$$\begin{cases} \left(\frac{\partial\rho}{\partial p}\right)_T \frac{\partial p}{\partial t} + v\left(\frac{\partial\rho}{\partial p}\right)_T \frac{\partial p}{\partial x} + \rho\frac{\partial v}{\partial x} + \left(\frac{\partial\rho}{\partial T}\right)_p \frac{\partial T}{\partial t} + v\left(\frac{\partial\rho}{\partial T}\right)_p \frac{\partial T}{\partial x} = 0 \\ v\left(\frac{\partial\rho}{\partial p}\right)_T \frac{\partial p}{\partial t} + \left[1 + v^2\left(\frac{\partial\rho}{\partial p}\right)_T\right]\frac{\partial p}{\partial x} + \rho\frac{\partial v}{\partial t} + 2\rho v\frac{\partial v}{\partial x} + v\left(\frac{\partial\rho}{\partial T}\right)_p \frac{\partial T}{\partial t} \\ \quad + v^2\left(\frac{\partial\rho}{\partial T}\right)_p \frac{\partial T}{\partial x} = J_2 \\ p\frac{\partial v}{\partial x} + \rho C_v\frac{\partial T}{\partial t} + \rho v C_v\frac{\partial T}{\partial x} = J_3 \\ \frac{\partial p}{\partial t} + \frac{dx}{dt}\cdot\frac{\partial p}{\partial x} = J_4 \\ \frac{\partial v}{\partial t} + \frac{dx}{dt}\cdot\frac{\partial v}{\partial x} = J_5 \\ \frac{\partial T}{\partial t} + \frac{dx}{dt}\cdot\frac{\partial T}{\partial x} = J_6 \end{cases}$$

where $J_2 = -\lambda\rho v\cdot|v|/2d_0 - \rho g\sin\alpha$; $J_3 = -4\kappa(T - T_{ex})/d_0 + \lambda\rho v^3/2d_0$; $J_4 = (dp/dt)_{x(t)}$; $J_5 = (dv/dt)_{x(t)}$; $J_6 = (dT/dt)_{x(t)}$. The total derivatives of the functions p, v and T are taken along the line $x(t)$, on which their values are known.

If the principal determinant of the system is non-vanishing, all six partial derivatives as well as the functions p, v, T along the curve $x(t)$ can be determined uniquely and independently from each other. If this determinant vanishes and the system of linear equations is compatible, the dependence between the values of p, v, T on the curve $x(t)$ exists.

Let us equate the principal determinant of the system of six linear equations to zero

$$\begin{vmatrix} \left(\frac{\partial\rho}{\partial p}\right)_T & v\left(\frac{\partial\rho}{\partial p}\right)_T & 0 & \rho & \left(\frac{\partial\rho}{\partial T}\right)_p & v\left(\frac{\partial\rho}{\partial T}\right)_p \\ v\left(\frac{\partial\rho}{\partial p}\right)_T & 1 + v^2\left(\frac{\partial\rho}{\partial p}\right)_T & \rho & 2\rho v & v\left(\frac{\partial\rho}{\partial T}\right)_p & v^2\left(\frac{\partial\rho}{\partial T}\right)_p \\ 0 & 0 & 0 & p & \rho C_v & \rho v C_v \\ 1 & dx/dt & 0 & 0 & 0 & 0 \\ 0 & 0 & 1 & dx/dt & 0 & 0 \\ 0 & 0 & 0 & 0 & 1 & dx/dt \end{vmatrix} = 0$$

and calculate it. To do this, let us multiply the first, third and fifth columns of this determinant by dx/dt and then subtract the resulting products from the second, fourth and sixth columns of the same determinant

$$\begin{vmatrix} \left(\frac{\partial \rho}{\partial p}\right)_T & (v-\dot{x})\left(\frac{\partial \rho}{\partial p}\right)_T & 0 & \rho & \left(\frac{\partial \rho}{\partial T}\right)_p & \left(\frac{\partial \rho}{\partial T}\right)_p (v-\dot{x}) \\ v\left(\frac{\partial \rho}{\partial p}\right)_T & 1+v(v-\dot{x})\left(\frac{\partial \rho}{\partial p}\right)_T & \rho & \rho(2v-\dot{x}) & v\left(\frac{\partial \rho}{\partial T}\right)_p & \left(\frac{\partial \rho}{\partial T}\right)_p v(v-\dot{x}) \\ 0 & 0 & 0 & p & \rho C_v & \rho C_v (v-\dot{x}) \\ 1 & 0 & 0 & 0 & 0 & 0 \\ 0 & 0 & 1 & 0 & 0 & 0 \\ 0 & 0 & 0 & 0 & 1 & 0 \end{vmatrix}$$

The obtained determinant can be calculated by the method of determinant decomposition in terms of the third-column elements. As a result we get the cubic equation

$$\rho C_v \left(\frac{dx}{dt} - v\right)^3 \left(\frac{\partial \rho}{\partial p}\right)_T + \left[\frac{p}{\rho}\left(\frac{\partial \rho}{\partial T}\right)_p - \rho C_v\right]\left(\frac{dx}{dt} - v\right) = 0 \tag{5.34}$$

with respect to the difference $(dx/dt - v)$. Roots of this equation are evident

1. $\frac{dx}{dt} - v = 0 \Rightarrow \frac{dx}{dt} = v$;
2. $\frac{dx}{dt} - v = \pm\sqrt{\frac{C_v - p/\rho^2 \cdot (\partial \rho/\partial T)_p}{C_v}\left(\frac{\partial p}{\partial \rho}\right)_T}$.

Since $C_v = \frac{\partial e_{in}}{\partial T}$ and $-\frac{p}{\rho^2}\left(\frac{\partial \rho}{\partial T}\right)_p = \frac{\partial}{\partial T}\left(\frac{p}{\rho}\right)_p$, then

$$C_v - \frac{p}{\rho^2}\left(\frac{\partial \rho}{\partial T}\right)_p = \frac{\partial}{\partial T}\left(e_{in} + \frac{p}{\rho}\right)_p = \left(\frac{\partial J}{\partial T}\right)_p = C_p$$

where $J(p, T)$ is enthalpy. The expression under the square root is simplified to

$$\frac{dx}{dt} - v = \pm\sqrt{\frac{C_p}{C_v}\left(\frac{\partial p}{\partial \rho}\right)_T} = \pm c$$

where

$$c^2 = \frac{C_p}{C_v} \cdot \left(\frac{\partial p}{\partial \rho}\right)_T = \gamma \cdot \left(\frac{\partial p}{\partial \rho}\right)_T$$

and $\gamma(p, T) = C_p/C_v$ is the adiabatic index.

The quantity c having the dimension of velocity is called the *adiabatic velocity of sound* in gas. For a perfect gas $(\partial p/\partial \rho)_T = RT$, $\gamma = \text{const.}$, $c = \sqrt{\gamma RT}$. For example, at $\gamma = 1.31$; $R = 450\ \text{J kg}^{-1}\ \text{K}^{-1}$; $T = 273$ K the velocity of sound is equal to $c = \sqrt{1.31 \cdot 450 \cdot 273} \cong 400\ \text{m s}^{-1}$.

The equation (5.34) provides *three* families of characteristics, two of them having velocities (slopes) $(dx/dt)_{1,2} = v \pm c$ and the third $dx/dt = v$. Since the gas velocity v in gas-pipelines as a rule does not exceed $10\ \mathrm{m\,s^{-1}}$ and $c \approx 400\ \mathrm{m\,s^{-1}}$, then sometimes we could take $dx/dt \cong \pm c$. We shall see below that the slope of characteristics to the t-axis is simply the velocity of propagation of *small perturbations* in gas (sound velocity). Therefore the assumption $dx/dt \cong \pm c$ means that this velocity of gas moving in a pipe is approximately equal to the sound velocity in quiescent gas. In the general case it is of course not so, especially for gas flowing with high velocity, for example in propulsive nozzles or for gas flowing from openings, orifices, nozzles and so on. In such cases the difference $(dx/dt - c)$ cannot be ignored.

Since the determinant of the system of the six linear equations under consideration vanishes at $\dot{x} = v$ and $\dot{x} = v \pm c$, for compatibility of this system the determinant

$$\begin{vmatrix} \left(\frac{\partial\rho}{\partial p}\right)_T & v\left(\frac{\partial\rho}{\partial p}\right)_T & 0 & \rho & \left(\frac{\partial\rho}{\partial T}\right)_p & 0 \\ v\left(\frac{\partial\rho}{\partial p}\right)_T & 1+v^2\left(\frac{\partial\rho}{\partial p}\right)_T & \rho & 2\rho v & v\left(\frac{\partial\rho}{\partial T}\right)_p & J_2 \\ 0 & 0 & 0 & p & \rho C_v & J_3 \\ 1 & \dot{x} & 0 & 0 & 0 & J_4 \\ 0 & 0 & 1 & \dot{x} & 0 & J_5 \\ 0 & 0 & 0 & 0 & 1 & J_6 \end{vmatrix}$$

obtained from the principle determinant of the system by replacement of the last column on free terms of the system of equations, should also vanish. In the theory of linear equations the condition of the latter determinant vanishing is called the *compatibility condition*. This condition as applied to our case may be called the *compatibility condition at characteristics*.

Characteristic form of equations. Omitting cumbersome calculation of this determinant, we give the final result

$$\left(\frac{dx}{dt} - v\right)^2 \left(\frac{\partial\rho}{\partial p}\right)_T \left(\frac{J_3}{\rho} - C_v J_6\right) - \left(\frac{dx}{dt} - v\right) \frac{p}{\rho^2} \left(\frac{\partial\rho}{\partial p}\right)_T (\rho J_5 - J_2)$$
$$- \left[\frac{J_3}{\rho} - C_p J_6 + \frac{p}{\rho^2}\left(\frac{\partial\rho}{\partial p}\right)_T J_4\right] = 0. \tag{5.35}$$

Substitution successively in Eq. (5.35) of the values of dx/dt for each of the characteristic families yields conditions for the functions p, v, T to be obeyed at these characteristics:

1. $$\frac{dp}{dt} + \rho c \frac{dv}{dt} = +c \cdot J_2 + \frac{(\gamma - 1)}{p/\rho \cdot (\partial\rho/\partial p)_T} \cdot J_3 \tag{5.36}$$

at $dx/dt - v = +c$ or $dx/dt = v + c$, where $J_2 = -\lambda\rho v \cdot |v|/2d_0 - \rho g \sin\alpha$, $J_3 = -4\kappa(T - T')/d_0 + \lambda\rho v^3/2d_0$ and derivatives with respect to time are taken in the direction of the characteristic $dx/dt = v + c$.

$$2.\ \frac{dp}{dt} - \rho c \frac{dv}{dt} = -c \cdot J_2 + \frac{(\gamma - 1)}{p/\rho \cdot (\partial\rho/\partial p)_T} \cdot J_3 \tag{5.37}$$

at $dx/dt - v = -c$ or $dx/dt = v - c$, where derivatives with respect to time are taken in the direction of the characteristic $dx/dt = v - c$.

$$3.\ C_p \frac{dT}{dt} - \frac{p}{\rho^2}\left(\frac{\partial\rho}{\partial p}\right)_T \frac{dp}{dt} = \frac{J_3}{\rho} \tag{5.38}$$

at $dx/dt - v = 0$ or $dx/dt = v$, where derivatives with respect to time are taken in the direction of the characteristic $dx/dt = v$.

Method of Characteristics

Formulas (5.36)–(5.38) give a constructive way to solve the system of equations (5.41). The method by which this solution is obtained is called the *method of characteristics.*

Basic equations for this method are Eqs. (5.36)–(5.38). The idea of this method is that at each point $M(x_k, t_m)$ of the plane (x, t) relevant to the time t_m "in the past", that is at $t < t_m$, there are just three characteristics $dx/dt = v \pm c$ and $dx/dt = v$, at which should be satisfied the compatibility conditions (5.36)–(5.38). Each of these compatibility conditions represents an ordinary differential equation which could be integrated over the direction of the respective characteristic. Thus, at the point $M(x_k, t_m)$, being the intersection point of these characteristics, appear three equations to determine three unknown quantities $p(x_k, t_m)$, $v(x_k, t_m)$ and $T(x_k, t_m)$. For numerical realization of the method of characteristics various schemes may be used. Consider one of them.

Let it be required to get the solution of the system of equations (5.33) in the region $0 < x < L, t > 0$ of the plane (x, t). Divide this region by a rectangular mesh with straight lines $x_{\hat{e}} = \Delta x \cdot (k - 1)$, $t_m = \Delta t \cdot (m - 1)$, where $1 \leq \kappa \leq N + 1$, $N = [L/\Delta x]$. The time step Δt is chosen so that $\Delta t = \Delta x/(v + c)_{\max}$, where $v_{\max}$ and $c_{\max}$ are the maximum possible values of the gas and sound velocities respectively ($c_{\max} = \sqrt{\gamma R T_{\max}}$; $T_{\max}$ is the maximum possible value of the gas temperature), $m = 1, 2, 3, \ldots$.

Neglecting the work of the gravity force, the basic system of differential equations (5.33) may be written in the so-called *divergent form* (see Eq. (1.36) at $\alpha_k \cong 1$)

$$\begin{cases} \dfrac{\partial\rho}{\partial t} + \dfrac{\partial\rho v}{\partial x} = 0 \\ \dfrac{\partial\rho v}{\partial t} + \dfrac{\partial}{\partial x}(p + \rho v^2) = -\lambda\rho v \cdot |v|/2d_0 \\ \dfrac{\partial}{\partial t}\left[\rho\left(\dfrac{v^2}{2} + e_{\text{in}}\right)\right] + \dfrac{\partial}{\partial x}\left[\rho v \cdot \left(\dfrac{v^2}{2} + e_{\text{in}} + \dfrac{p}{\rho}\right)\right] = -\dfrac{4\kappa}{d_0}(T - T_{\text{ex}}) \end{cases} \tag{5.39}$$

The left-hand sides of Eqs. (5.39) are represented by differential operators of the form

$$\operatorname{div}\{A, B\} = \frac{\partial A}{\partial t} + \frac{\partial B}{\partial x}$$

expressing the *divergence* of a vector with coordinates $\{A, B\}$ in the space of variables (x, t).

Let us first integrate the system of equations (5.39) over the area of the mesh cell ABCD ($x_k \le x \le x_{k+1}; t_{m-1} \le t \le t_m$) with sides Δx and Δt (Fig. 5.11) and transform the integrals over the mesh area into integrals over the mesh contour using the formula

$$\iint_{ABCD} \left(\frac{\partial A}{\partial t} + \frac{\partial B}{\partial x} \right) \mathrm{d}x\mathrm{d}t = \oint_{ABCD} [A\cos(nt) + B\cos(nx)]\mathrm{d}\sigma$$

$$= \oint_{ABCD} (A\mathrm{d}x + B\mathrm{d}t) = (A_{k+1/2,m} - A_{k+1/2,m-1})\Delta x + (B_{k+1} - B_k)\Delta t.$$

Here the quantities with fraction subscripts denote mean values of the corresponding parameters at the horizontal sides (BC and AD) of the mesh cell, whereas the quantities with integer subscripts denote mean values of parameters at the vertical sides (AB and CD) of the same mesh cell.

Application of this transformation to each equation of the system (5.39) yields the following system of finite difference equations

$$\rho_{k+1/2,m} = \rho_{k+1/2,m-1} - [(\rho v)_{k+1} - (\rho v)_k] \cdot \frac{\Delta t}{\Delta x},$$

$$\rho_{k+1/2,m} v_{k+1/2,m} = \rho_{k+1/2,m-1} v_{k+1/2,m-1} - \left[(p + \rho v^2)_{k+1} - (p + \rho v^2)_k \right]$$
$$\cdot \frac{\Delta t}{\Delta x} - (\lambda \rho v \cdot |v|/2d_0)_{k+1/2,m-1} \cdot \Delta t,$$

$$\rho_{k+1/2,m} \left(\frac{v^2}{2} + C_v T \right)_{k+1/2,m} = \rho_{k+1/2,m} \left(\frac{v^2}{2} + C_v T \right)_{k+1/2,m-1}$$
$$- \left\{ \left[\partial v \cdot \left(\frac{v^2}{2} + C_v T + \frac{\rho}{\partial} \right) \right]_{k+1} \right.$$

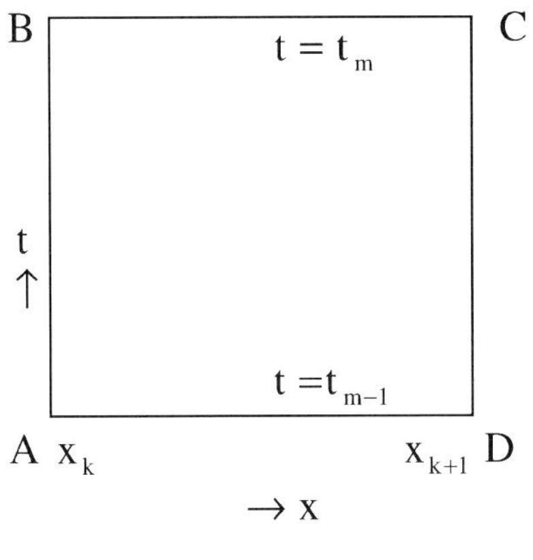

Figure 5.11 Mesh cell in the plane of variables (x, t).

$$- \left[\partial v \cdot \left(\frac{v^2}{2} + C_v T + \frac{\rho}{\partial}\right)\right]_k \Bigg\}$$

$$\times \frac{\Delta t}{\Delta x} - \frac{4\kappa}{d_0}(T - T_{ex})_{k+1/2,m-1} \cdot \Delta t. \tag{5.40}$$

Here $\rho_{k+1/2,m}$; $v_{k+1/2,m}$ and $T_{k+1/2,m}$ are, respectively, the mean values of the density, velocity and temperature of the gas in segment (x_k, x_{k+1}) of the gas-pipeline at the instant of time $t = t_m$; $\rho_{k+1/2,m-1}$, $v_{k+1/2,m-1}$ and $T_{k+1/2,m-1}$ are the mean values of the same functions at the previous moment of time $t_{m-1} = t_m - \Delta t$.

The physical meaning of the obtained relations is clear: each of these equations represents the integral balance of one or other parameter in the mesh cell. For example, the first relation (5.40) reflects the fact that the gas mass $\rho_{k+1/2,m} \cdot \Delta x$ in the segment (x_k, x_{k+1}) of the pipeline at the instant of time t_m is equal to the gas mass $\rho_{k+1/2,m-1} \cdot \Delta x$ in the same segment at the previous instant of time added to the mass difference of the gas $[(\rho v)_{k+1} - (\rho v)_k] \cdot \Delta t$ flowing in time Δt from the kth segment into the $(k+1)$th segment and from the $(k-1)$th segment into the kth one. The two other relations are interpreted in the same manner with the only difference being that they deal with momentum and total energy, respectively.

Nevertheless, Eqs. (5.40) are not closed since they include unknown quantities denoted by integer subscripts and representing the transfer of mass, momentum and energy from one mesh cell into another. The essence of the method is that the values of these quantities are found from compatible conditions (5.36)–(5.38) at characteristics.

Let us represent the conditions (5.32)–(5.37) in the form of finite difference equations

$$\frac{p_{k,m} - p_{k-1/2,m-1}}{\Delta t} + (\rho c)_{k-1/2,m-1}\frac{v_{k,m} - v_{k-1/2,m-1}}{\Delta t} = \left(c \cdot J_2 + \frac{(\gamma-1)}{p/\rho \cdot (\partial\rho/\partial p)_T} \cdot J_3\right)_{k-1/2,m-1},$$

$$\frac{p_{k,m} - p_{k+1/2,m-1}}{\Delta t} - (\rho c)_{k+1/2,m-1}\frac{v_{k,m} - v_{k+1/2,m-1}}{\Delta t} = \left(-c \cdot J_2 + \frac{(\gamma-1)}{p/\rho \cdot (\partial\rho/\partial p)_T} \cdot J_3\right)_{k+1/2,m-1}, \tag{5.41}$$

$$C_p\frac{T_{k,m} - T_{k-1/2,m-1}}{\Delta t} - \left[\frac{p}{\rho^2}\left(\frac{\partial\rho}{\partial p}\right)_T\right]_{k-1/2,m-1} \cdot \frac{p_{k,m} - p_{k-1/2,m-1}}{\Delta t} = \left(\frac{J_3}{\rho}\right)_{k-1/2,m-1}.$$

As a result we have three linear equations for three unknown quantities $p_{k,m}, v_{k,m}, T_{k,m}$. First we find quantities $p_{k,m}$ and $v_{k,m}$ from the first two equations, then from the third equation we obtain $T_{k,m}$ and from the relation $\rho_{k,m} = p_{k,m}/RT_{k,m}$ we calculate the gas density. Substituting $\rho_{k,m}, p_{k,m}, v_{k,m}, T_{k,m}$ into Eq. (5.40), we get average values of the gas-dynamic parameters in the considered mesh cell at the instant of time t_m.

To Eqs. (5.37) and (5.38) should be added initial and boundary conditions. Boundary conditions vary depending on the concrete problem. As an example, they could be taken in the form of two relations between the pressure, flow rate and temperature of the gas at one edge (left) of the pipeline section, reflecting conditions of compressor station operation and one condition at the right edge of the pipeline section, for example, with a given pressure.

5.5
Gas Outflow from a Pipeline in the Case of a Complete Break of the Pipeline

Let us illustrate the use of the method of characteristics as applied to non-stationary processes in a gas-pipeline with an example of the calculation of gas outflow from a pipeline in the case of its complete break. The dynamics of gas outflow happen as a rule in two regimes. At first at the cross-section, through which occurs gas outflow, appears a critical regime with the velocity of gas outflow equal to the local velocity of sound ($\approx$380–400 m s^{-1}). After the pressure in the gas-pipeline is lowered by a certain value (for natural gas by a factor of 1.8–1.9 greater than atmospheric pressure) the outflow regime becomes subsonic and the gas velocity gradually decreases from sound velocity to zero.

The process of gas outflow is not isothermal. The gas temperature, owing to adiabatic expansion and the Joule-Thomson effect falls significantly at the break in the cross-section as well as far from it. For example, with a complete break of a gas-pipeline the gas temperature can be reduced by 80–100 K. Only at the final stage of the process is there a gradual restoration of temperature due to external heat inflow.

Numerical calculations are carried out on the basis of recurrent formulas (5.38). The left-hand end of the pipeline ($x = 0$) is closed, whereas the right-hand end ($x = L$) is suddenly opened and remains open to the atmosphere.

Hence, the boundary conditions are as follows.

1. *Cross-section* $x = 0$. The first boundary condition is $v_{1,m}(0, t_m) = 0$, since the left-hand end of the pipeline is taken to be closed. Hence it follows that at $v = 0$ the left integration boundary coincides with one of the characteristics of the differential equations (5.36), therefore the second boundary condition at the cross-section $x = 0$ ($k = 1$) is formulated as a condition at the characteristic $\mathrm{d}x/\mathrm{d}t = v = 0$

$$C_p \frac{T_{1,m} - T_{1,m-1}}{\Delta t} - \left[\frac{p}{\rho^2} \left(\frac{\partial \rho}{\partial p} \right)_T \right]_{1,m-1} \cdot \frac{p_{1,m} - p_{1,m-1}}{\Delta t} = \left(\frac{J_3}{\rho} \right)_{1,m-1}.$$

The compatibility condition at the characteristic $dx/dt = v - c$ coming from the integration domain to the initial pipeline cross-section (in our case at the characteristic $dx/dt = 0 - c = -c$), provides the third boundary condition

$$\frac{p_{1,m} - p_{3/2,m-1}}{\Delta t} - (\rho c)_{3/2,m-1} \frac{v_{1,m} - v_{3/2,m-1}}{\Delta t} = \left(-c \cdot J_2 + \frac{(\gamma - 1)}{p/\rho \cdot (\partial \rho / \partial p)_T} \cdot J_3 \right)_{3/2,m-1}.$$

2. *Cross-section* $x = L$. Two characteristics of the system (5.39) come to this cross-section from the integration domain. These characteristics are $dx/dt = v + c$ and $dx/dt = v$ with positive slope. Thus, the first two boundary conditions at $x = L$ $(k = N + 1)$ are

$$\frac{p_{N+1,m} - p_{N+1/2,m-1}}{\Delta t} + (\rho c)_{N+1/2,m-1} \frac{v_{N+1,m} - v_{N+1/2,m-1}}{\Delta t} = \left(c \cdot J_2 + \frac{(\gamma - 1)}{p/\rho \cdot (\partial \rho / \partial p)_T} \cdot J_3 \right)_{N+1/2,m-1},$$

$$C_p \frac{T_{N+1,m} - T_{N+1/2,m-1}}{\Delta t} - \left[\frac{p}{\rho^2} \left(\frac{\partial \rho}{\partial p} \right)_T \right]_{N+1/2,m-1} \cdot \frac{p_{N+1,m} - p_{N+1/2,m-1}}{\Delta t} = \left(\frac{J_3}{\rho} \right)_{N+1/2,m-1}.$$

The third boundary condition at $x = L$ has a different form depending on whether the gas outflow is subsonic $v < c(p, T)$ or sonic $v = c(p, T)$. If the gas outflow happens with local sound velocity, then $dx/dt = v - c = 0$ and the right boundary of the integration domain coincides with the characteristic of the system (5.39). Hence, the third boundary condition is nothing but a condition at this characteristic

$$\frac{p_{N+1,m} - p_{N+1,m-1}}{\Delta t} - (\rho c)_{N+1,m-1} \frac{v_{N+1,m} - v_{N+1,m-1}}{\Delta t} = \left(-c \cdot J_2 + \frac{(\gamma - 1)}{p/\rho \cdot (\partial \rho / \partial p)_T} \cdot J_3 \right)_{N+1,m-1}.$$

In subsonic flow $v < c(p, T)$ only two characteristics of the system (5.39) come from the integration domain to the boundary points $x = L$, therefore one more condition should be given. Such a condition is $p_{N+1,m} = p_{\text{aTM}}$ meaning that the pressure at the opened pipeline end is equal to the external atmospheric pressure.

Calculations show that the gas is significantly cooled during outflow. Figure 5.12 shows graphs of the gas temperature distribution over a pipeline

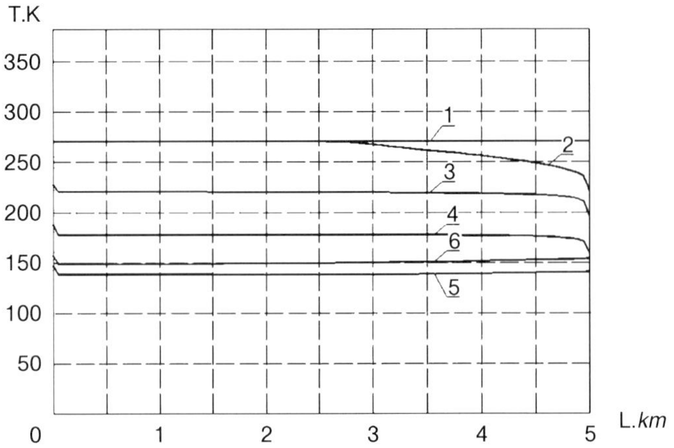

Figure 5.12 Distribution of gas temperature over a pipeline section length: 1, $t = 0$; 2, $t = 5$ s; 3, $t = 60$ s; 4, $t = 120$ s; 5, $t = 200$ s; 6, $t = 420$ s.

section length ($D = 1220$ mm, $L = 5$ km) at different instants of time. It is seen that the initial temperature of the gas, equal to $0\,^{\circ}$C decreases by more than 100 K. This is explained by gas expansion due to the high outflow velocity and the Joule–Thomson effect. Gas cooling in the pipeline happens because the above-mentioned effects are too fast to be compensated by heat inflow from the surrounding medium.

Computer modeling shows the effect of gas suction into the pipeline in the final stage of the outflow process. Figure 5.13 represents the dependence of the gas velocity on time at the cross-section where the pipeline undergoes a break ($D = 1220$ mm, $L = 1$ km) with initial pressure $p(x, 0) = 5.5$ MPa. The oscillation process is seen to appear in the final stage of gas outflow and lasts for about 2 min.

The section AB of the graph characterizes sonic outflow of gas lasting about 20 s. The velocity of the outflow gradually decreases as a consequence of reduction in the pressure and temperature of the gas at the cross-section of the break. The section BF of the graph characterizes subsonic outflow of gas. About after a 30th of a second from the outflow beginning in the pipeline an oscillation process occurs in which the gas periodically changes its direction of motion. Once the velocity of the gas becomes equal to zero it continues to decrease and at the section CD of the graph remains negative, testifying that atmospheric air is being sucking into the pipeline. The same behavior of gas is observed in the following instants of time. Calculations show that the maximal velocity of air suction in the oscillation process exceeds 50 m s^{-1}.

Since the mixture of natural gas and air could achieve an explosive concentration, the discovered phenomena appears to be a serious hazard to attendants. In particular, it is strongly recommended to take the utmost care in repair-reconditioning operations.

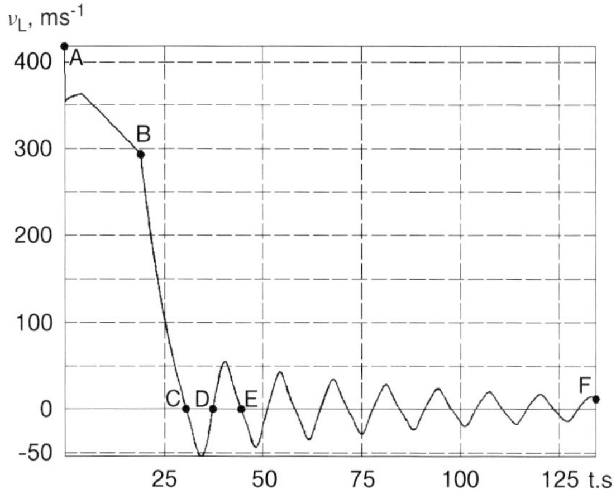

Figure 5.13 Dependence of gas outflow velocity on time.

5.6 Mathematical Model of Non-Stationary Gravity Fluid Flow

Such a flow has already been discussed in Section 3.7 but then a *stationary* gravity flow was considered, that is a flow in which all the hydrodynamic parameters remain constant at each cross-section of the pipeline. In other words these parameters were independent of time. Consider now a one-dimensional mathematical model of *non-stationary* gravity fluid flow. In this model there are two parameters governing the flow: $S(x, t)$ the area of the pipeline cross-section and $v(x, t)$ the flow velocity (Fig. 5.14).

The differential equations of the model under consideration are

$$\begin{cases} \dfrac{\partial \rho S}{\partial t} + \dfrac{\partial \rho v S}{\partial x} = 0 \\ \dfrac{\partial \rho v S}{\partial t} + \dfrac{\partial \rho v^2 S}{\partial x} = -\rho g S \cos\alpha \cdot \dfrac{\partial h}{\partial x} - \rho g S \sin\alpha - \dfrac{\rho g S \cos\alpha}{C_{\mathrm{Sh}}{}^2 R_{\mathrm{h}}} \cdot v|v|. \end{cases}$$

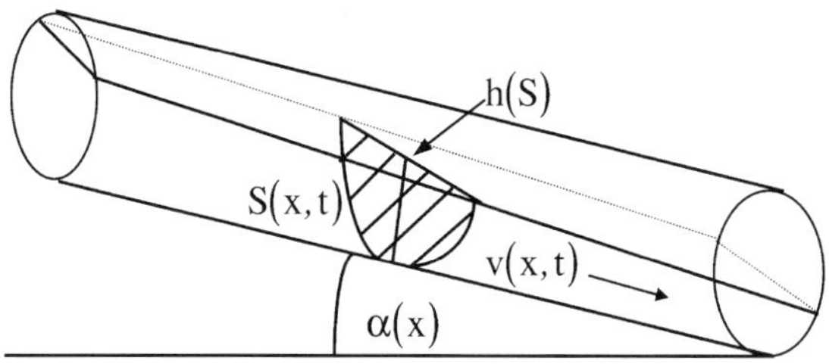

Figure 5.14 Gravity fluid flow.

There $R_h(S)$ is the hydraulic radius of the flow (see Section 3.7); $h(S)$ is the depth of the pipeline cross-section filling with fluid; α is the angle of inclination of the pipeline axis to the horizontal; C_{Sh} is the Chezy factor.

The first of these equations (continuity equation) expresses the law of mass conservation in the fluid flow in a pipeline with cross-section partially filled by fluid. The second equation (momentum equation) is the law of momentum change, that is Newton's Second law: on the left is the derivative of momentum with respect to time, the quantity $\rho v^2 S = \rho v S \cdot v$ representing the flux of momentum; the term $\rho g S \cos\alpha \cdot \frac{\partial h}{\partial x}$ on the right is the Boussinesq force acting on the fluid due to its free surface being non-parallel to the pipeline axis, i.e. due to excess of the fluid level at one cross-section of the pipeline as compared to the fluid level at another cross-section; the term $-\rho g S \sin\alpha$ is the component of the gravity force and the term $-\rho g S \cos\alpha \cdot v|v|/({C_{Sh}}^2 R_h)$ is the force of the flow resistance due to fluid friction against the pipeline walls.

The system of equations may be rewritten in an equivalent form if we take into account that

$$\rho g \cos\alpha \cdot \int_0^S S dh = \rho g \cos\alpha \cdot \int_0^S S \frac{dh(S)}{dS} dS$$

As a result we get

$$\begin{cases} \dfrac{\partial \rho S}{\partial t} + \dfrac{\partial \rho v S}{\partial x} = 0 \\ \dfrac{\partial \rho v S}{\partial t} + \dfrac{\partial}{\partial x}\left(\rho v^2 S + \rho g \cos\alpha \cdot \displaystyle\int_0^S S dh\right) \\ \quad = -\rho g S \sin\alpha - \dfrac{\rho g S \cos\alpha}{{C_{Sh}}^2 R_h} \cdot v|v| \end{cases} \tag{5.42}$$

The system of equations (5.42) belongs to the class of *quasi-linear (that is linear with respect to derivatives) differential equations of the hyperbolic type*. Solution of these equations can be obtained by specially elaborated methods, which are beyond the scope of this book. We advise those who are interested in these methods to consult the book by Rozhdestvenski and Yanenko (1977), which among other things contains voluminous literature devoted to this problem.

For gravity stationary fluid flow ($\partial/\partial t = 0$) from the first equation of the system (5.42) it follows that $\rho v S =$ const. If we take into account that $\rho \cong \rho_0 =$ const., then $v \cdot S =$ const. If we assume in addition that $S =$ const., which is true everywhere except in small regions close to the transfer sections where gravity flows are formed, from Eq. (5.42) it follows that

$$0 = -\rho g S \cdot \sin\alpha - \frac{\rho g S \cos\alpha}{{C_{Sh}}^2 R_h} \cdot v|v| \Rightarrow v = C_{Sh}\sqrt{R_h \cdot \tan\alpha} \tag{5.43}$$

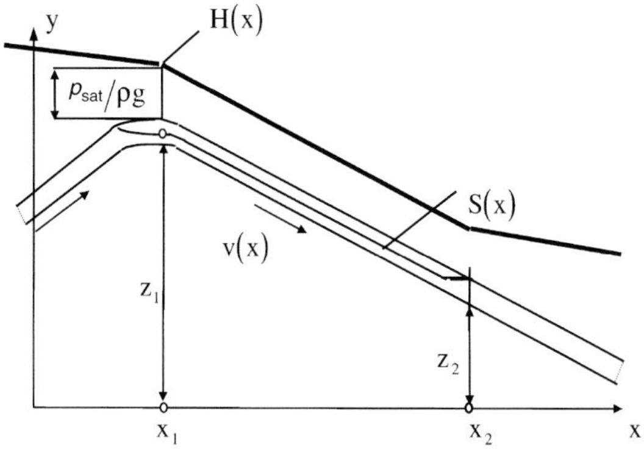

Figure 5.15 Gravity section in pipeline.

which after some transformation allows one to get the degree of filling with fluid of the gravity flow section, σ, depending on the hydraulic gradient $\tan\alpha$ (see Eq. (3.51)).

The point at which the first gravity section in the pipeline begins is called the *transfer section*. Since cross-sections after the transfer section are only partially filled by fluid, the pressure in this section is constant and equal to the saturated vapor pressure p_{sat} of this fluid. Figure 5.15 illustrates the behavior of the hydraulic gradient in gravity flow. It is seen from this figure that in this section the line of the hydraulic gradient goes parallel to the pipeline axis at a distance $p_{sat}/\rho g$ from it owing to the pressure constancy in the gas cavity over the fluid-free surface.

An ordinary differential equation serves to calculate the depth $h(x)$ of fluid in the pipe.

$$\frac{dh}{dx} = -\frac{\tan\alpha + Q^2/(C_{Sh}{}^2 R_h S^2)}{1 - (Q^2 dh/dS)/(g\cos\alpha \cdot S^3)} \tag{5.44}$$

following from Eq. (5.42) for the case of stationary flow ($\partial/\partial t = 0$) with $h(x) = h[S(x)]$; $R_h = R_h(S)$; Q is the fluid flow rate.

The depth h_n of the fluid in the pipe at which the numerator of the fraction on the right-hand side of Eq. (5.44) vanishes is called the *normal depth* of gravity flow in the pipe. Fluid flow with a normal depth of flow happens under the condition $h_n = \text{const}$. In this case the relation (5.43) holds.

The depth h_{cr}, at which the nominator of the fraction on the right-hand side of Eq. (5.44) vanishes is called the *critical depth*. In cross-sections with such a depth the derivative dh/dx tends to ∞ and the fluid flow varies the level of pipe filling abruptly. Depending on the relation between the depths h_n and h_{cr} different flow regimes are possible. Investigations of these regimes are described in special monographs (Archangelskiy, 1947; Leibenson *et al.*, 1934; Christianovitch, 1938).

5.7
Non-Stationary Fluid Flow with Flow Discontinuities in a Pipeline

The preceding classical models of non-stationary flows of a slightly compressible fluid (see Sections 5.1–5.3) contained an essential restriction on the absence of phase transition in the fluid. It was tacitly supposed that the fluid under no circumstances passes into the vapor-gas phase, even when the pressure drops to the value of the saturated vapor pressure. However, in the propagation of a rarefaction wave in the pipeline this condition can be violated at many pipeline cross-sections and first and foremost at the tops of the pipeline profile. When the pressure in the rarefaction wave reduces to the saturated vapor pressure the fluid boils, the vapor column breaks and the pipeline cross-section becomes partially filled with vapor. From this point on all further results predicted by classical theory appear to be wrong.

For example, at the disconnection of a pumping station or an aggregate of this station a rarefaction wave is propagated downstream in the pipeline. The pressure in such a wave reduces leading to the formation of voids at the tops of the pipeline profile. These voids are capable of growing and turning into stationary gravity flow sections or, on the contrary, contracting and even disappearing altogether. To perform calculation of such processes on the base of classical theory is of course impossible.

We can give another example: closing the gate valve generates a compression wave propagating upstream with a rise in the pressure in the wave. This wave having been reflected from the open surface of the reservoir or from a vapor-gas cavity inside the pipeline initiates a rarefaction wave traveling in the opposite direction and reduces the pressure in the fluid. This brings into existence temporal transfer points at some tops of the pipe profile and cavities. This leads sometimes to fluid flows with a partially filled pipeline cross-section. If the pressure supply in the pipeline is not high, the reduction in pressure leads to flow discontinuity and the generation of vapor-gas cavities. So, for example, in laboratory installations it could be observed that the fluid before the gate valve literally boiled owing to the sharp reduction in pressure. Calculation in these cases on the basis of classical theory is also impossible.

One more example can be given: Connection of a lateral tap from the oil-pipeline to an intermediate oil tank leads to the propagation of rarefaction waves up and down stream from the place of tap cutting up- and downstream. These waves are able to break the fluid column at many cross-sections of the pipeline profile and turn the enforced flow into a gravity one characterized by the presence of vapor-gas cavities and gravity flow sections. Such cases also defy calculation in the framework of classical theory.

All the aforesaid is true also for pipelines transporting the so-called *unstable fluids*, among which are gas-condensate and a wide fraction of light hydrocarbons with saturated vapor pressure from 3 to 30 atm. Any sharp pressure reduction gives rise to a plethora of vapor-gas cavities the disappearance of which in the pipe lead to powerful hydraulic shocks.

Profile Hydraulic Shock

It should be noted that the appearance or disappearance of voids in the pipeline is unsafe and is rather dangerous for pipeline integrity. In particular, in Lurie and Polyanskaya (2000), the discovery and investigation of the origin of powerful hydraulic waves in the pipeline called profile hydraulic shock are described. For example, each time the gate valve, located before a section of pipeline with significant slope, was closed the rarefaction resulting in the region close to the valve led to a sequence of powerful hydraulic shocks. These shocks were gradually damped with time and the flow in the pipeline was stabilized. Similar phenomena were observed in the pipeline on disconnection of some pumps or the pumping station as a whole.

The nature of the profile hydraulic shock is as follows. When the gate valve (at $x = 0$, Fig. 5.16) closes the pressure in the sloping section of pipeline falls and the fluid column, initially supported by this pressure, begins little by little to slip down and acquire inverse flow (Figs. 5.17–5.19). The fluid in this inverse flow is accelerated and when vapor-gas voids, having originated before the gate valve, are taken out, there is an abrupt stop of fluid flow and, as a consequence, hydraulic shock (Fig. 5.18). The power of the shock is then especially great, since

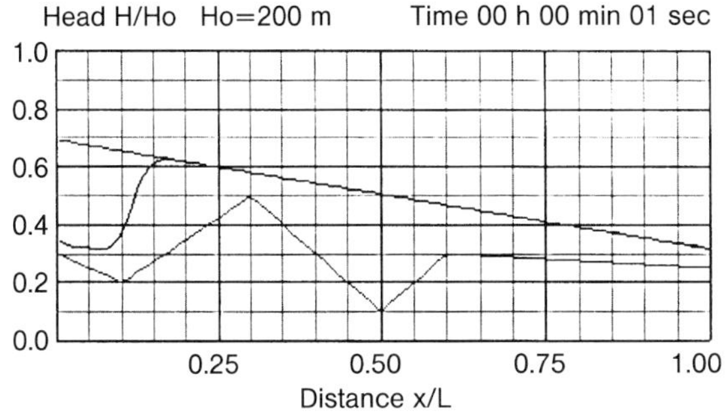

Figure 5.16 Initial stage of the process.

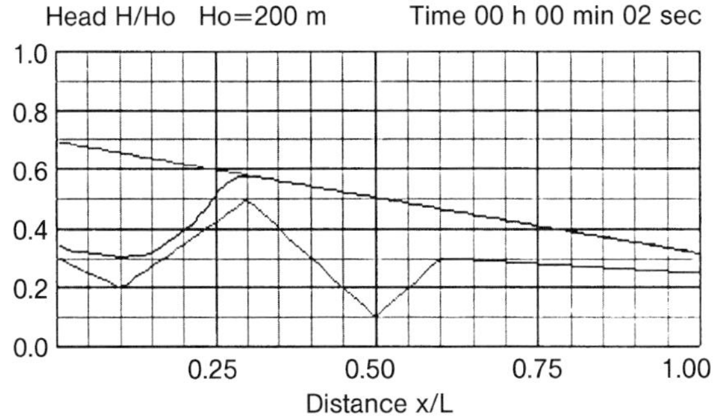

Figure 5.17 Formation of gravity flow in the upward sloping pipeline section.

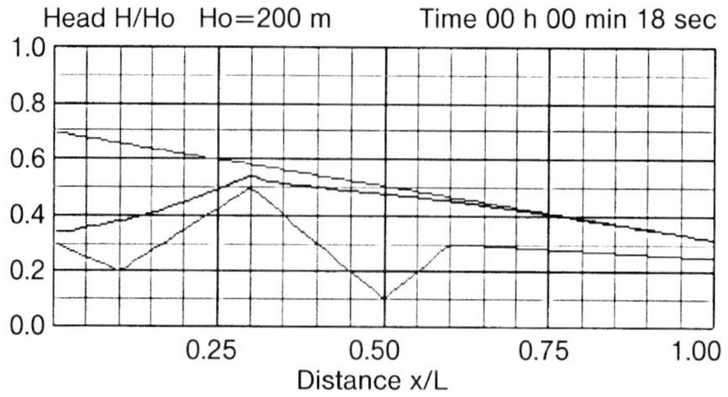

Figure 5.18 Formation of reverse fluid flow in the upward sloping pipeline section.

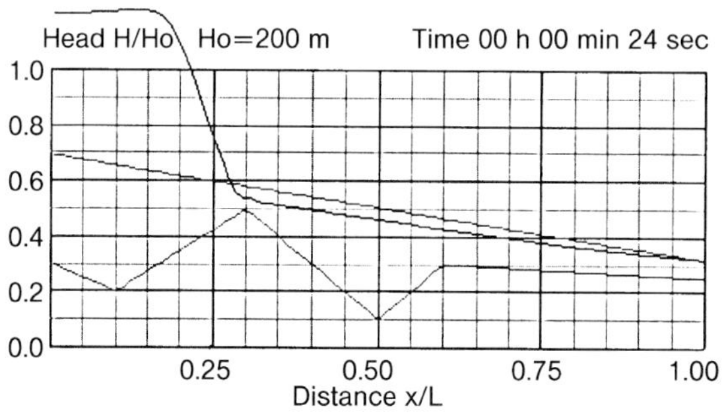

Figure 5.19 Profile hydraulic shock.

the inverse flow of the fluid runs against the closed gate valve and, as follows from the Joukovski formulas, the amplitude of the hydraulic shock redoubles.

The hydraulic shock wave reflected from the gate valve and accompanied by fluid stop, propagates downstream. Reaching a temporary transfer section, at which a vapor-gas void has been formed, the wave is reflected from this void and now, in the form of a rarefaction wave, heads back to the gate valve, the fluid column in the ascending section of the pipeline again becomes weightless and begins to slip down to the closed gate valve. Then a secondary hydraulic shock occurs, after which the process is repeated again and again with decreasing intensity.

Generalized theory, (Lurie and Polyanskaya, 2000). In accordance with the classical theory of non-stationary processes the wave processes generating in a completely filled pipeline at its start-up or stopping, opening or closing of the gate valve or lateral tap and so on, are described by differential equations (5.4) for pressure $p(x, t)$ and velocity $v(x, t)$

$$\begin{cases} \dfrac{\partial p}{\partial t} + \rho_0 c^2 \cdot \dfrac{\partial v}{\partial x} = 0 \\ \rho_0 \dfrac{\partial v}{\partial t} + \dfrac{\partial p}{\partial x} = -\lambda(Re, \varepsilon)\dfrac{1}{d_0}\dfrac{\rho_0 v|v|}{2} - \rho_0 g \sin\alpha(x) \end{cases}$$

These equations represent the laws of mass conservation and variation of momentum of fluid particles moving in a pipeline. In the generalized theory it is assumed that in the pipeline there are completely filled enforced (pumped) sections as well as only partially filled gravity sections in which the pressure is equal to the saturated vapor pressure p_{sat}. The fluid flow in the sections of enforced flow is described by the system of equations (5.4) whereas in the sections of gravity flow it is described by Eqs. (5.42).

Method of calculation. For numerical calculation of oil product flow in completely filled as well as in partially filled pipeline sections there is an elaborate scheme of *end-to-end calculation* based on the ideas of Godunov (Samarskiy, 1977). This scheme involves consideration of the so-called *problems on the disintegration of arbitrary discontinuity* in the system of hyperbolic equations.

Results of calculation. Figures 5.16–5.23 demonstrate the results of calculations on successive stages of unloading wave propagation in a 10-km pipeline with internal diameter $d = 516$ mm at an abrupt drop in the pumping delivery of benzene ($\rho = 750$ kg m^{-3}, $p_{sat} = 0.7$ atm) from 1500 to 200 m^3 h^{-1}. It is seen that, as distinct from existing theory, the line of hydraulic gradient at

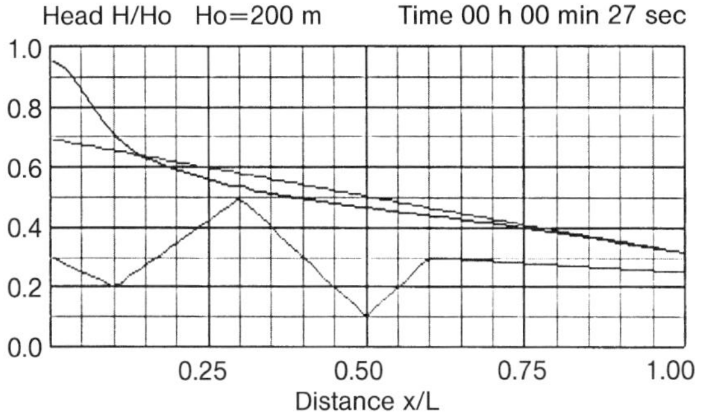

Figure 5.20 Decay of hydraulic shock wave.

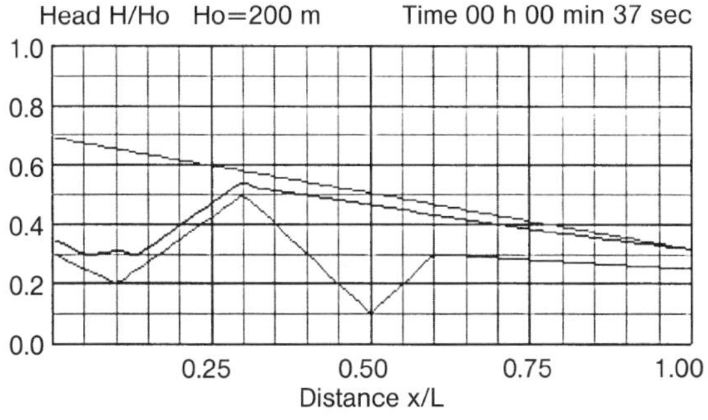

Figure 5.21 Formation of the secondary reverse flow in the upward sloping pipeline section.

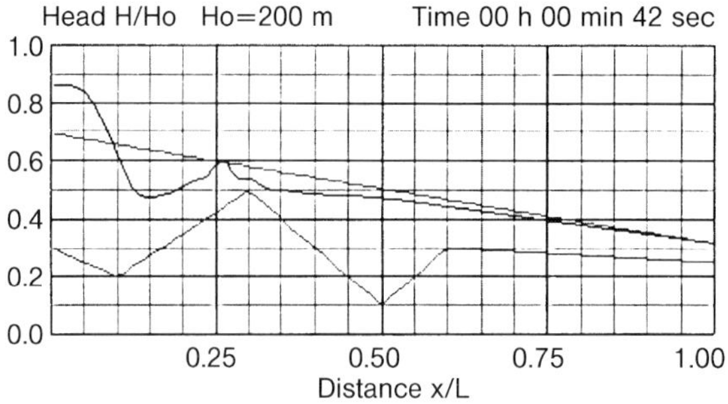

Figure 5.22 The second top of the profile hydraulic shock.

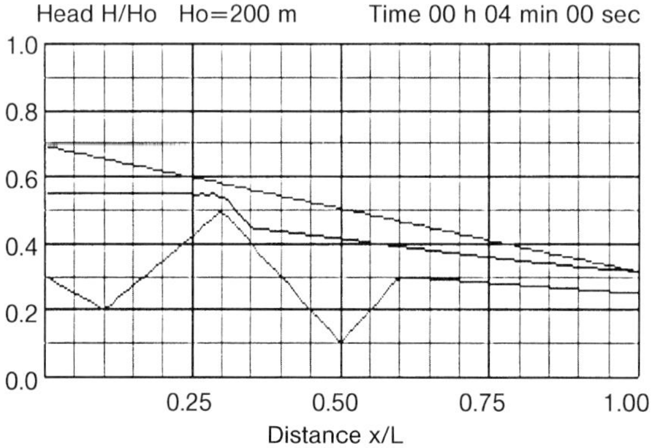

Figure 5.23 Formation of new stationary flow.

no instant of time intersects the pipeline profile, that is, the pressure in the pipeline would never be less than the fluid saturated vapor pressure.

The figures also show how the wave of pressure decrease runs over the pipeline profile top (lower polygonal line of Fig. 5.16), and how, at this top, a temporary transfer point and further gravity flow section with inverse flow of fluid is being formed (Fig. 5.17 and Fig. 5.18).

Since the fluid flows in the opposite direction (as if it would run on a closed gate valve), in 24 s after the aggregate disconnection a powerful hydraulic shock is formed in the pipe (Fig. 5.19), in which the pressure is nearly twice the initial pressure at the station. The amplitude of the hydraulic shock wave gradually decreases (Fig. 5.20) and in a further 10 s in the upward sloping section of the pipeline slope the inverse flow appears again (Fig. 5.21). After 46 s the secondary hydraulic shock occurs (Fig. 5.22), but already with lesser amplitude. In the given calculation there were six such hydraulic shocks. Only 4 min after the pumping regime change the stationary regime in the pipeline is achieved (Fig. 5.23), where there exists a single gravity flow section of length 500 000 m in the pipeline downward sloping section.

6 Dimensional Theory

Dimensional theory contains fundamental propositions for representing equations of mathematical regularities modeling different phenomena in *invariant form*, that is, in a form independent of the choice of the units of the measurements. Such representations permit one to compile classes of similar phenomena and to model them on experimental installations.

When setting out the fundamentals of dimensional and similarity theory and the modeling of phenomena we have followed the methodology of Sedov (1965).

6.1 Dimensional and Dimensionless Quantities

A quantitative description of various physical phenomena, including the transport of fluids and gases in pipelines, is connected with measurements of the characteristics of these phenomena, whose numerical values *depend on the choice of measurement units*. For example, pipeline diameter can be expressed through the numbers 1; 10; 100; 1000; 3.28; 39.4 and so on depending on that what unit of measurement is taken: meter, decimeter, centimeter, millimeter, foot or inch (1 foot $\cong$ 0.3048 m; 1 inch $\cong$ 0.0254 m); the length of a pipeline can be expressed through the numbers 150 000; 150 or 93.21 and so on, depending on what is used as the measurement unit of the length: meter, kilometer or mile (1 mile $\cong$ 1.6093 km). The same is true for many other physical quantities. For example, the volumetric fluid flow rate in a pipeline may be measured by the numbers 1000, 277.8 or 73.4 depending on whether the units of volume and time are taken as cubic meter and hour; liter and second; gallon (1 gallon in USA $\cong$ 3.785 liter) and second. The pressure can also be measured with different numbers 64, 6.28 or 0.0433 depending on the measurement units: technical atmosphere (kgf s^{-1} m^{-2}), millions of pascals (megapascals, MPa) or psi (pound per square inch: 1 force pound $\cong$ 4.448 N; 1 mass pound $\cong$ 453.6 g, 1 psi $\cong$ 0.006896 MPa $\cong$ 0.0703 atm).

It is important to note that the choice of measurement units *depends on the researcher* and therefore is, to a great extent, arbitrary. Two different researchers

Modeling of Oil Product and Gas Pipeline Transportation. Michael V. Lurie

ISBN: 978-3-527-40833-7

describing one and the same phenomenon but at the same time using different measurement units can obtain diverse numerical values of one and the same parameter.

And yet there exist parameters whose numerical values *do not depend on the researcher*, that is, do not depend on measurement units. These parameters are said to be *invariant* relative to the choice of measurement units. For example, the ratio of the pipeline length to the pipeline diameter L/d, the ratio of the pressure at a pumping station discharge line to the pressure at a pumping station suction line p_{dl}/p_{sl} or a more complex combination such as the Reynolds number $Re = vd/\nu$ (v is the average fluid velocity in the pipeline, d is the internal diameter of the pipeline, ν is the fluid kinematic viscosity) are independent of the choice of measurement units. This means that the numerical values of L/d, p_{dl}/p_{sl} and $v \cdot d/\nu$ would be unchanged for any choice of length, pressure, velocity and viscosity units.

Hence, it is possible to give the following definition: *quantities whose numerical values depend on the choice of measurement units are called dimensional quantities; quantities whose numerical values do not depend on the choice of measurement units are called dimensionless quantities.*

6.2
Primary (Basic) and Secondary (Derived) Measurement Units

Measurement units, having been introduced empirically by arbitrary conditions and propositions, are called primary or basic units. Among these are, in particular, the units of length, time and mass. In the international SI system they are defined as follows.

Meter is a unit of length measurement. In accordance with the definition taken at the 11th General Conference on Weights and Measures (1960), 1 meter is a length equal to 1650763.73 lengths of the wave emitted by the krypton (Kr) atom in vacuum at its energy-level transition. The international standard of the meter before 1960 was a bar of platinum–iridium alloy marked on one side of its planes. This bar is kept on deposit at the International Bureau of Weights and Measures in Sevre near Paris. At first the meter was defined as 10^{-7} part of a quarter of the Earth's meridian.

Second is the measurement unit of time. There are recognized the atomic (standard) second reproduced by cesium standards of frequency and time and the ephemeris second equal to 1/31556925.9747 part of the tropical year.

Kilogram is a unit of mass measurement. The standard kilogram is equal to the mass of the international prototype kept at the International Bureau of Weights and Measures. The prototype of the kilogram is made from platinum–iridium alloy in the form of a cylindrical weight. The relative error of standard copies compared with the original does not exceed $2 \cdot 10^{-9}$.

There are also other basic units of measurement, such as coulomb (C) – a unit of electricity quantity (electrical charge), volt (V) – a unit of electrical stress (voltage, potential); degrees ($^\circ$C, $^\circ$F, K and other) – a unit of temperature and so on.

Measurement units of other quantities are by definition introduced through the basic measurement units. Such units are called *secondary or derived units.* Among them are the following.

Velocity which is defined as the ratio of length to time, therefore velocity units can be m s^{-1}, km h^{-1}, mile h^{-1} and so on.

Acceleration which is defined as the ratio between the velocity increment and time, hence acceleration units can be m s^{-2}, foot s^{-2}, km s^{-2}, mile h^{-2} and so on.

Force (weight) which is defined as the product of mass and acceleration, thus force units can be dyne (1 dyn = 1 g cm s^{-2}); newton (1 N = 1 kg m s^{-2}); pound ($\cong$4.448 kg m s^{-2}) and so on.

Density which is defined as the mass of the medium unit volume, therefore its units can be kg m^{-3}, g cm^{-3}, t m^{-3} and so on.

Specific weight which is defined as the weight of the medium unit volume, hence its measurement units can be N m^{-3}, dyn cm^{-3} and so on.

Pressure which is defined as the ratio of force to unit area, thus force units can be: pascal (1 Pa = 1 N m^{-2} = 1 kg m^{-1} s^{-2}), pound $inch^{-2}$ and so on.

Fluid flow rate (e.g. volumetric) which is defined as the fluid volume crossing the pipeline cross-section area in a unit time, therefore its units can be m^3 s^{-1}, m^3 h^{-1}, m^3 min^{-1}, l s^{-1} and so on.

Current intensity which is defined as a charge transmitted in a unit time, hence the unit of measurement can be the ampere (1 A = 1 C s^{-1}).

All these and analogous measurement units are derived from the basic units.

6.3
Dimensionality of Quantities. Dimensional Formula

Let there be a physical parameter A. Expression of its measurement unit through basic units is called the *dimension* of a given parameter and is denoted by the symbol $[A]$. This expression written as a formula is called the *dimensional formula.*

Denote through L the measurement unit of length, through T the measurement unit of time and through M the measurement unit of mass. Then the expression of the measurement units of many other quantities can be written as the following formulas:

$$(A = v) \text{ – velocity: } [v] = \frac{\mathrm{L}}{\mathrm{T}} = \mathrm{L} \cdot \mathrm{T}^{-1} = \mathrm{M}^0 \cdot \mathrm{L} \cdot \mathrm{T}^{-1};$$

$$(A = a) \text{ – acceleration: } [a] = \frac{\mathrm{L}}{\mathrm{T}^2} = \mathrm{L} \cdot \mathrm{T}^{-2} = \mathrm{M}^0 \cdot \mathrm{L} \cdot \mathrm{T}^{-2};$$

$$(A = F) \text{ – force: } [F] = \mathrm{M} \cdot [\mathrm{a}] = \mathrm{M} \cdot \mathrm{L} \cdot \mathrm{T}^{-2};$$

$$(A = \rho) \text{ – density: } [\rho] = \frac{\mathrm{M}}{\mathrm{L}^3} = \mathrm{M} \cdot \mathrm{L}^{-3} \cdot \mathrm{T}^0;$$

$$(A = \gamma) - \text{specific weight } [\gamma] = \frac{\mathrm{M \cdot L \cdot T^{-2}}}{\mathrm{L^3}} = \mathrm{M \cdot L^{-2} \cdot T^{-2}};$$

$$(A = p) - \text{pressure: } [p] = [\mathrm{F}]/\mathrm{L}^2 = \frac{\mathrm{M \cdot L \cdot T^{-2}}}{\mathrm{L^2}} = \mathrm{M \cdot L^{-1} \cdot T^{-2}};$$

$$(A = Q) - \text{volumetric flow rate: } [Q] = \frac{\mathrm{L^3}}{\mathrm{T}} = \mathrm{M^0 \cdot L^3 \cdot T^{-1}};$$

$$(A = G) - \text{mass flow rate: } [G] = \frac{\mathrm{M}}{\mathrm{T}} = \mathrm{M^1 \cdot L^0 \cdot T^{-1}};$$

$$(A = B) - \text{dimensionless parameter } [B] = \mathrm{M^0 \cdot L^0 \cdot T^0}.$$

Thus, particular examples show that in all cases the dimensional formula of parameter A has form of a power monomial

$$[A] = \mathrm{M}^{m_1} \cdot \mathrm{L}^{m_2} \cdot \mathrm{T}^{m_3} \tag{6.1}$$

where m_1, m_2, m_3 are certain real positive or negative numbers.

What does the dimensional formula mean? It allows one to determine very simply *by how many-fold the numerical value of parameter A would be changed, on going from one system of basic measurement units to another one differing from the first system only by the scales of the basic units.* For example, if the transition from the new system of basic measurement units to the old one is accomplished by variation of the mass unit k_1-fold, of the length unit k_2-fold, of the time unit k_3-fold, the numerical value of the parameter A would vary $k_1^{m_1} \cdot k_2^{m_2} \cdot k_3^{m_3}$-fold, that is the new value $\grave{A}'$ of this parameter would be determined by the formula

$$\grave{A}' = k_1^{m_1} \cdot k_2^{m_2} \cdot k_3^{m_3} \cdot \grave{A}. \tag{6.2}$$

Let us explain the afore said by examples.

Example 1. The velocity v in the SI system of basic measurement units (m, s, kg) is 1 m s^{-1}. For a new system of basic measurement units (cm, s, kg) the transition from this system to the SI system is performed by increasing one of the basic units (length) by a factor of 100, the value of the velocity will also be enhanced 100 times

$$v' = 1^0 \cdot 100^1 \cdot 1^{-1} = 100 \cdot v = 100 \text{ cm s}^{-1}.$$

Example 2. The pressure p in the SI system of basic units (m, s, kg) is 6 MPa, that is 6 000 000 kg m^{-1} s^{-2}. For a new system of units (inch, s, pound) the transition from this system to the SI system is carried out by changing the length scales $100/2.54 = 39.37$ times and the mass by $1/0.4536 = 2.205$ times, hence the value of the pressure will vary by a factor $2.205^1 \cdot 39.37^{-1} \cdot 1^{-2} = 0.056$ and will be $0.056 \cdot 6\,000\,000 = 336\,043$ (psi).

In Eq. (6.1), obtained empirically, there are only three basic units of measurement because we have considered examples solely from mechanics. However, it is easy to verify that even if the number of physical parameters is increased by other parameters from electrical, heat, chemical and other phenomena and the number of basic units grows and exceeds three, the dimensional formula would not be radically changed. Only the number of factors will increase. Hence, it could be assumed, that the dimensional formula in the general case would have the form of a power monomial

$$[A] = [a_1]^{m_1} \cdot [a_2]^{m_2} \cdot [a_3]^{m_3} \cdot \ldots \cdot [a_n]^{m_n} \tag{6.3}$$

where A is a physical parameter, whose dimension is derived from the dimensions of basic quantities denoted by $a_1, a_2, \ldots, a_n$.

6.4 Proof of Dimensional Formula

Let us prove the validity of the dimensional formula (6.3) in the general case. This formula means that if we change the scale of a basic measurement unit a_1 by k_1 times, of a_2 by k_2 times, ..., of a_n by k_n times, the numerical value of the parameter A would be changed by $k_1^{m_1} \cdot k_2^{m_2} \cdot \ldots \cdot k_n^{m_n}$ times, that is the new value of this parameter A' would be equal to

$$A' = k_1^{m_1} \cdot k_2^{m_2} \cdot \ldots \cdot k_n^{m_n} \cdot A.$$

Let us use this circumstance.

Let there be three researchers B, C and D measuring one and the same physical parameter A but using different systems of basic measurement units, namely

B: $\{a_1, a_2, \ldots, a_n\}$;

C: $\{a'_1, a'_2, \ldots, a'_n\}$, so that its basic units are related to the basic units of researcher B by the formulas

$$\begin{aligned} a'_1 &= \alpha_1 \cdot a_1 \\ a'_2 &= \alpha_2 \cdot a_2 \\ &\ldots\ldots\ldots\ldots \\ a'_n &= \alpha_n \cdot a_n \end{aligned}$$

where $k_1 = \alpha_1, k_2 = \alpha_2, \ldots, k_n = \alpha_n$;

D: $\{a''_1, a''_2, \ldots, a''_n\}$, so that its basic units are related to the basic units of researcher C by the formulas

$$\begin{aligned} a''_1 &= \beta_1 \cdot a'_1 \\ a''_2 &= \beta_2 \cdot a'_2 \\ &\ldots\ldots\ldots\ldots \\ a''_n &= \beta_n \cdot a'_n \end{aligned}$$

where $k_1 = \beta_1, k_2 = \beta_2, \ldots, k_n = \beta_n$, and to the basic units of researcher B by the formulas

$$a_1'' = \beta_1\alpha_1 \cdot a_1$$
$$a_2'' = \beta_2\alpha_2 \cdot a_2$$
$$\ldots\ldots\ldots\ldots$$
$$a_n'' = \beta_n\alpha_n \cdot a_n$$

where $k_1 = \beta_1\alpha_1, k_2 = \beta_2\alpha_2, \ldots, k_n = \beta_n\alpha_n$.

It is reasonable to suppose that the values of the parameter A measured by the three researchers would be different: A measured by B, $\grave{A}'$ measured by C and $\grave{A}''$ measured by D.

Let a function $F(k_1, k_2, \ldots, k_n)$ show how many-fold the numerical value of parameter A in one of the system of basic units would be changed when passing to another system differing by $k_1, k_2, \ldots, k_n$ times from the first one in the scales of the basic units, respectively. Then if we go from the units of researcher B to the units of researcher C, we obtain

$$A' = F(\alpha_1, \alpha_2, \ldots, \alpha_n) \cdot A.$$

If we then go from the units of researcher C to the units of researcher D, we get

$$\grave{A}'' = F(\beta_1, \beta_2, \ldots, \beta_n) \cdot A'$$

or

$$\grave{A}'' = F(\beta_1, \beta_2, \ldots, \beta_n) \cdot F(\alpha_1, \alpha_2, \ldots, \alpha_n) \cdot A. \tag{6.4}$$

On the other hand, if we go at once from the units of researcher B to the units of the researcher D, bypassing researcher C, we get

$$\grave{A}'' = F(\alpha_1\beta_1, \alpha_2\beta_2, \ldots, \alpha_n\beta_n) \cdot A \tag{6.5}$$

The result should of course be independent of the transition route from one researcher to another one, thus it must *identically* satisfy the following functional equation

$$F(\alpha_1\beta_1, \alpha_2\beta_2, \ldots, \alpha_n\beta_n) = F(\alpha_1, \alpha_2, \ldots, \alpha_n) \cdot F(\beta_1, \beta_2, \ldots, \beta_n) \tag{6.6}$$

which ought to be valid for *any* values of the factors $\alpha_1, \alpha_2, \ldots, \alpha_n$ and $\beta_1, \beta_2, \ldots, \beta_n$.

Find the solution of this equation.

Differentiation of both parts of the identity (6.6) with respect to β_i, where i may be equal to $1, 2, 3, \ldots, n$, yields

$$\alpha_i \frac{\partial F}{\partial \xi_i} = \frac{\partial F}{\partial \beta_i} \cdot F(\alpha_1, \alpha_2, \ldots, \alpha_n), \text{ where } \xi_i = \alpha_i\beta_i.$$

Since this equality also represents an identity, then $\beta_1 = 1, \beta_2 = 1, \ldots, \beta_n = 1$. As a result the following differential equation is obtained

$$\alpha_i \cdot \frac{\partial F(\alpha_1, \alpha_2, \ldots, \alpha_n)}{\partial \alpha_i} = m_i \cdot F(\alpha_1, \alpha_2, \ldots, \alpha_n) \tag{6.7}$$

in which $m_i = \partial F/\partial \beta_i$ at $\beta_1 = 1, \beta_2 = 1, \ldots, \beta_n = 1$.

The solution of differential equation (6.7) gives the dependence of the sought function F on parameter α_i. Really, the separation of variables provides

$$\frac{\partial}{\partial \alpha_i}(\ln F) = \frac{m_i}{\alpha_i} \Rightarrow \ln F = m_i \cdot \ln \alpha_i + c \Rightarrow F = K \cdot \alpha_i^{m_1}$$

where the integration constant K ($K = \mathrm{e}^C$) is a function of the remaining parameters α_j. Since α_i was any one of the arguments of the function F, the latter should have the following form

$$F = K_0 \cdot \alpha_1^{m_1} \cdot \alpha_2^{m_2} \cdot \ldots \cdot \alpha_n^{m_n}$$

where $K_0 = \text{const}$. If $F(1, 1, \ldots, 1) = 1$, because the value of the parameter A is not changed by variation of the basic measurement unit scales, $K_0 = 1$ and the function will be

$$F = \alpha_1^{m_1} \cdot \alpha_2^{m_2} \cdot \ldots \cdot \alpha_n^{m_n}$$

or redesignating the variables as $k_1, k_2, \ldots, k_n$, one gets

$$F = k_1^{m_1} \cdot k_2^{m_2} \cdot \ldots \cdot, k_n^{m_n}. \tag{6.8}$$

Hence, when changing the basic unity a_1 by k_1 times, the numerical value of the parameter A would vary by ${k_1}^{m_1}$ times, when changing the basic unity a_2 by k_2 times the numerical value of the parameter A would vary by ${k_2}^{m_2}$ and so on. Thus, the parameter A has the dimension

$$[A] = [a_1]^{m_1} \cdot [a_2]^{m_2} \cdot [a_3]^{m_3} \cdot \ldots \cdot [a_n]^{m_n}$$

which proves the dimensional formula (6.3).

6.5 Central Theorem of Dimensional Theory

It is appropriate now to interrupt the description of the theory and to formulate a question concerning an apparent contradiction due to the use of dimensional quantities.

Let the mathematical form of a certain physical phenomenon be expressed by the dependence of a parameter A on other parameters $a_1, a_2, \ldots, a_n$ governing this phenomenon as follows

$$A = f(a_1, a_2, \ldots, a_n). \tag{6.9}$$

For example, the enhancement of pipeline cross-section area ΔS when producing a positive pressure P depends on the magnitude of this pressure as well as on the pipeline diameter D, the wall thickness δ and the elastic modulus (Young's modulus) E of the steel from which the pipeline is made. Hence we have $A = \Delta S$, $a_1 = P$, $a_2 = D$, $a_3 = \delta$, $a_4 = E$, so that $\Delta S = f(P, D, \delta, E)$.

It is evident that the dependence under consideration exists objectively and should not depend either on the researcher performing the investigation or on the choice of measurement units being used to calculate the values of the function, i.e. parameter A, and its arguments $(a_1, a_2, \ldots, a_n)$.

On the other hand all quantities entering into the dependence (6.9) are dimensional quantities whose numerical values depend on the choice of measurement units and consequently on the researcher.

The question arises as *to at which point the dependence (6.9) reflects the objectively existing physical regularity when numerical values of the function and its arguments depend on the researcher.*

The answer to this question gives the central theorem of dimensional theory called the *Buckingham* **Π** *theorem*. This theorem states: "*Each dependence between dimensional quantities reflecting an objectively existing physical regularity could be rewritten in invariant form independent of the choice of measurement units, namely in the form of a dependence between dimensional complexes composed from governing parameters*".

Now we will prove this theorem, removing the contradiction between the objective character of any physical regularity and the subjective character of the choice of measurement units. However, before we do this, we need first to define dimensionally-dependent and dimensionally-independent quantities.

6.6
Dimensionally-Dependent and Dimensionally-Independent Quantities

It is said that a quantity "a" is dimensionally-dependent on the quantities $a_1, a_2, \ldots, a_n$ when its dimension $[a]$ is expressed through the dimensions $[a_1], [a_2], \ldots, [a_n]$ by the formula

$$[a] = [a_1]^{m_1}[a_2]^{m_2}\ldots[a_n]^{m_n} \tag{6.10}$$

namely there exist such numbers $m_1, m_2, \ldots, m_n$ that the equality (6.10) is obeyed. If such numbers do not exist, it is said that the quantity a is dimensionally-dependent on the quantities $a_1, a_2, \ldots, a_n$.

It is evident that parameters with dimensions of time t, length l and mass m are dimensionally-independent of each other. Parameters with dimensions of velocity v and density ρ are dimensionally-independent. On the other hand, a parameter with the dimension of pressure p is dimensionally-dependent on parameters with dimensions of density ρ and velocity v. It is easy to verify that

$[p] = [\rho][v]^2$

$$[p] = \mathrm{M}^1\mathrm{L}^{-1}\mathrm{T}^{-2}; [\rho] = \mathrm{M}^1\mathrm{L}^{-3}\mathrm{T}^0; [v^2] = \mathrm{M}^0\mathrm{L}^2\mathrm{T}^{-2} \Rightarrow$$

$$\mathrm{M}^1\mathrm{L}^{-1}\mathrm{T}^{-2} = (\mathrm{M}^1\mathrm{L}^{-3}\mathrm{T}^0)(\mathrm{M}^0\mathrm{L}^2\mathrm{T}^{-2}).$$

Hence, the ratio $p/(\rho v^2)$ is dimensionless since the dimensions of the numerator and denominator coincide.

There is a general algorithm capable of determining whether one or another parameter is dimensionally-dependent or dimensionally-independent of other given parameters. For example, let us consider mechanics, in which there are three basic measurement units: mass (M), length (L), time (T) and it is required to clarify whether the parameter P with the dimension of pressure is dimensionally-dependent on parameters μ with the dimension of dynamic viscosity, g with the dimension of acceleration and D with the dimension of length. Let us show how to do it.

Write the dimensions of all the parameters under consideration

$$[p] = \mathrm{M}^1\mathrm{L}^{-1}\mathrm{T}^{-2},$$

$$[\mu] = \mathrm{ML}^{-1}\mathrm{T}^{-1},$$

$$[g] = \mathrm{LT}^{-2},$$

$$[\mathrm{D}] = \mathrm{L}.$$

Now let us look for the numbers m_1, m_2 and m_3 that would obey the equality

$$[p] = [\mu]^{m_1}[g]^{m_2}[D]^{m_3}.$$

Insertion of the considered dimensions in this equality yields

$$\mathrm{M}^1\mathrm{L}^{-1}\mathrm{T}^{-2} = (\mathrm{ML}^{-1}\mathrm{T}^{-1})^{m_1}(\mathrm{LT}^{-2})^{m_2}(\mathrm{L})^{m_3}.$$

Equating the exponents of identical basic measurement units in the left- and right-hand sides of the last relation, we get a system of three linear equations

$$1 = m_1,$$

$$-1 = -m_1 + m_2 + m_3,$$

$$-2 = -m_1 - 2m_2$$

to determine the three unknown quantities m_1, m_2, m_3. This system has a single solution

$$m_1 = 1, \ m_2 = \frac{1}{2}, \ m_3 = -\frac{1}{2}.$$

Thus, we obtain

$$[p] = [\mu][g]^{1/2}[D]^{-1/2} \tag{6.11}$$

which proves that pressure is a dimensionally-dependent parameter in the system of three parameters – viscosity, acceleration, length. From Eq. (6.11) it follows that the ratio $p/(\mu g^{1/2} D^{-1/2})$ is a dimensionless quantity since the dimensions of the numerator and denominator coincide.

In a similar manner many examples of the same type could be considered. Let us focus our attention on the question of the *maximal number of dimensionally-dependent quantities* for a given set of dimensional parameters $\{a_1, a_2, \ldots, a_n\}$. If this set contains n elements, it is always possible to separate from it a subset containing the maximum possible number $k \leq n$ of dimensionally-independent parameters.

This assertion follows from the known theorem of algebra that states that from any system of linear equations can be separated the maximum possible number of linearly independent equations, this number being called the *rank* of the system of equations. If we write dimensional formulas for all the parameters $a_1, a_2, \ldots, a_n$,

$$[a_1] = [O_1]^{x_1}[O_2]^{x_2} \ldots .[O_k]^{x_k},$$

$$[a_2] = [O_1]^{y_1}[O_2]^{y_2} \ldots .[O_k]^{y_k},$$

$$\ldots\ldots\ldots\ldots\ldots\ldots\ldots\ldots.,$$

$$[a_n] = [O_1]^{z_1}[O_2]^{z_2} \ldots .[O_k]^{z_k}$$

where $O_1, O_2, \ldots, O_k$ are symbols of basic measurement units in the given system of units and $x_1, x_2, \ldots, x_k; y_1, y_2, \ldots, y_k; z_1, z_2, \ldots, z_k$ are exponents of this formula, the question of the dimensional dependence (or independence) of each of them on the other as well as the question of the maximum number of dimensionally-independent quantities is reduced to solutions of the system of linear equations

$$m_1 x_1 + m_2 y_1 + \ldots + m_n z_1 = 0,$$

$$m_1 x_2 + m_2 y_2 + \ldots + m_n z_2 = 0,$$

$$\ldots\ldots\ldots\ldots\ldots\ldots\ldots\ldots\ldots\ldots$$

$$m_1 x_k + m_2 y_k + \ldots + m_n z_k = 0$$

(k equations with n unknowns) with respect to the unknowns $m_1, m_2, \ldots, m_n$ and to determine the rank of the system matrix.

In order to separate a subset containing the maximum number of dimensionally-independent quantities of the set $\{a_1, a_2, \ldots, a_n\}$ we proceed as follows. Let us take a quantity a_1. If it is a dimensional quantity, the next quantity a_2 is added to it. If a_2 has dimension different from the dimension of a_1, the system $\{a_1, a_2\}$ would consist of dimensionally-independent quantities. Next a_3 is added to the system of quantities $\{a_1, a_2\}$ a_3. If the dimension of a_3 is expressed through $\{a_1, a_2\}$ by the formula (6.10), this quantity is rejected and instead we take a_4, which in its turn is tested for independence from the quantities $\{a_1, a_2\}$. If the dimension of a_3 is not expressed through the dimensions of $\{a_1, a_2\}$, the system $\{a_1, a_2, a_3\}$ represents a system of dimensionally-independent quantities. In such a way all quantities in the set

$\{a_1, a_2, \ldots, a_n\}$ are considered and gradually the subset containing the number of dimensionally-independent quantities is separated.

Note that for one and the same set *several different subsets* containing the maximum number of dimensionally-independent quantities could be separated. Such subsets by themselves can be different but the number of elements in them would be equal.

In particular, when the dimensions of all quantities in the set $\{a_1, a_2, \ldots, a_n\}$ are expressed through the dimensions M, L, T, this set can have *no more than three* dimensionally-independent parameters.

Exercise. It is required to separate the maximum number of dimensionally-independent quantities among a set of the following parameters: p – pressure; ρ – density; v – velocity; Q – volumetric flow rate; ν – kinematic viscosity; D – diameter; g – acceleration due to gravity.

Solution. Write the dimensional formulas for all given parameters using the SI system of basic measurement units

$$[p] = \frac{\text{kg}}{\text{m} \times \text{s}^2}; \quad [\rho] = \frac{\text{kg}}{\text{m}^3}; \quad [v] = \frac{\text{m}}{\text{s}}; \quad [Q] = \frac{\text{m}^3}{\text{s}};$$

$$[\nu] = \frac{\text{m}^2}{\text{s}}; \quad [g] = \frac{\text{m}}{\text{s}^2}; \quad [D] = \text{m}.$$

Since all the parameters are expressed through mass, length and time, the maximum number of dimensionally-independent quantities is less than or equal to three. As the first quantity, we take D. As the second quantity we take g, because its dimension contains time and consequently it cannot be expressed through the dimensions of D. Finally, as the third dimensionally-independent quantity we take ρ. It is evident that the dimension of ρ cannot be expressed through the dimensions of D and g, since it contains mass. Hence, the set $\{D, g, \rho\}$ consisting of three parameters represents a basis of the maximum number of dimensionally-independent quantities in the given set of parameters.

There are of course other possible subsets of the given set consisting also of a maximum number of dimensionally-independent quantities. It is easy to verify that, for example, the subsets $\{p, v, D\}$, $\{\rho, Q, D\}$ and so on consist of dimensionally-independent quantities and the number of dimensionally-independent quantities in each of these subsets is a maximum and equal to three.

Exercises.

1. It is required to separate a basis consisting of the maximum number of dimensionally-independent quantities among a set of the following parameters: ρ – density; ω – frequency of revolutions; D – wheel diameter; Q – volumetric flow rate; g – acceleration due to gravity.

   Answer. For example $\{\rho, \omega, D\}$.

2. It is required to separate a basis consisting of a maximum number of dimensionally-independent quantities among a set of the following parameters: v – velocity; S – area of the cross-section occupied by fluid; g – acceleration due to gravity; D – diameter of the pipeline; α – angle of inclination of the pipeline axis to the horizontal (dimensional quantity); ν – kinematic viscosity, $m^2\ s^{-1}$.

   Answer. For example $\{D, v\}$.

3. It is required to separate a basis consisting of a maximum number of dimensionally-independent quantities among a set of the following parameters: p – pressure; v – velocity; θ – temperature; ρ – density; λ – thermal diffusivity factor, $W\ m^{-1}\ K^{-1}$; c – heat capacity, $J\ kg^{-1}\ K^{-1}$).

   Answer. For example $\{\rho, v, c, \theta\}$.

6.7 Buckingham Π-Theorem

Now we go to the proof of the central theorem of dimensional theory, the Buckingham Π-theorem, which was partially formulated in Section 6.5.

Let a physical regularity be represented by a function

$$A = f(a_1, a_2, \ldots, a_k, a_{k+1}, \ldots, a_n) \tag{6.12}$$

of arguments $a_1, a_2, \ldots, a_n$ among which can be dimensional as well as dimensionless parameters.

Let the maximum number of dimensionally-independent arguments in the set $a_1, a_2, \ldots, a_n$ be k and without disturbing the generality it can be taken that the first k arguments are $a_1, a_2, \ldots, a_k$. Then the remaining $(n-k)$ arguments of this function $a_{k+1}, a_{k+2}, \ldots, a_n$ would be dimensionally-dependent on the first k arguments, that is

$$[a_{k+1}] = [a_1]^{m_1}[a_2]^{m_2}\ldots[a_k]^{m_k},$$

$$[a_{k+2}] = [a_1]^{n_1}[a_2]^{n_2}\ldots[a_k]^{n_k},$$

$$\ldots\ldots\ldots\ldots\ldots\ldots\ldots\ldots$$

$$[a_n] = [a_1]^{p_1}[a_2]^{p_2}\ldots[a_k]^{p_k}$$

where $m_i, n_i, \ldots, p_i$ are real numbers. Thus, the relations

$$\Pi_1 = \frac{a_{k+1}}{a_1^{m_1}a_2^{m_2}\cdots a_k^{m_k}},$$

$$\Pi_2 = \frac{a_{k+2}}{a_1^{n_1}a_2^{n_2}\cdots a_k^{n_k}}, \tag{6.13}$$

$$\Pi_{n-k} = \frac{a_n}{a_1^{p_1}a_2^{p_2}\cdots a_k^{p_k}}$$

are dimensionless parameters since the dimensions of the quantities in the numerators and denominators of these fractions are identical.

The dimension of the quantity A should also be expressed through the dimensions of the arguments $a_1, a_2, \ldots, a_k$. If it is not expressed through these dimensions, it would also not be expressed through the dimensions of all the quantities $a_1, a_2, \ldots, a_n$. Hence, there should exist a relation

$$[A] = [a_1]^{q_1}[a_2]^{q_2} \ldots [a_k]^{q_k}$$

meaning that the ratio

$$\Pi = \frac{A}{a_1^{q_1} a_2^{q_2} \cdots a_k^{q_k}} \tag{6.14}$$

has to be dimensionless.

Let us revert to the dependence (6.12). This dependence may be rewritten as

$$\frac{A}{a_1^{q_1} a_2^{q_2} \cdots a_k^{q_k}} = \tilde{f}\left(a_1, a_2, \ldots, a_k, \frac{a_{k+1}}{a_1^{m_1} a_2^{m_2} \cdots a_k^{m_k}}, \ldots, \frac{a_n}{a_1^{p_i} a_2^{p_2} \cdots a_k^{p_k}}\right)$$

where $\tilde{f}$ represents a function resulting from f after redefinition of its arguments. This dependence with (6.13) and (6.14) can be represented as follows

$$\Pi = \tilde{f}(a_1, a_2, \ldots, a_k, \Pi_1, \Pi_2, \ldots, \Pi_{n-k}). \tag{6.15}$$

If we now arbitrarily and independently from each other vary the numerical values of the arguments $a_1, a_2, \ldots, a_k$ by going from one system of basic measurement units to another one, the numerical values of the parameters Π and $\Pi_1, \Pi_2, \ldots, \Pi_{n-k}$ would not be changed because they are dimensionless quantities. From here it follows that the function Π cannot depend on its first k arguments $a_1, a_2, \ldots, a_k$, since the dependence (6.15) would have the following simple form

$$\Pi = \tilde{f}(\Pi_1, \Pi_2, \ldots, \Pi_{n-k}). \tag{6.16}$$

Thus, we have shown that any physical dependence between dimensional quantities of the type (6.12) will be invariant, that is, independent of the choice of measurement units, of the form (6.16) between dimensionless complexes made up from arguments of the dependence under consideration. The number of these complexes would be less than the number of arguments of the initial dependence by a number equal to the maximum number of dimensionally-independent quantities among these arguments.

Exercise 1. The augmentation of the cross-section area ΔS of a steel pipeline when setting up in it excess pressure P depends on the value of this pressure, the diameter D of the pipeline, the thickness δ of the pipeline wall and the elastic modulus (Young's modulus) E of the steel from which the pipe is made.

Using the Π-theorem, it is required to write this dependence in dimensionless form and to clarify how many dimensionless parameters determine it.

Solution. The dependence to be sought can be written in a general form as $\Delta S = f(P, D, \delta, E)$. The dimensions of the arguments of this dependence in the SI system are: $[P] = \text{kg m}^{-1}\,\text{s}^{-2}$; $[D] = \text{m}$; $[\delta] = \text{m}$; $[E] = \text{kg m}^{-1}\,\text{s}^{-2}$. Among these there are only two dimensionally-independent quantities, for example D and E. Consequently, the number of arguments in the dependence under consideration may be reduced to two. Thus we have:

$$\Pi = S/D^2;\; \Pi_1 = P/E;\; \Pi_2 = \delta/D;$$

$$\Pi = \tilde{f}(\Pi_1, \Pi_2) \text{ or } \Delta S = D^2 \cdot \tilde{f}(P/E, \delta/D).$$

Hence, the dimensional analysis has shown that in the considered dependence there are only two dimensionless complexes P/E and δ/D instead of four dimensional arguments.

Moreover, if additionally we invoke the reasoning that the variation of pipeline cross-section area ΔS should be proportional to P/E (the function, $\tilde{f}$, could be expanded in a Taylor series in the vicinity of the point $\Pi_1 = 0$ leaving in the expansion only the first term, because $\Delta S = f(0, \Pi_2) = 0$ at $P = 0$ and the ratio P/E is very small: $P \approx 2–7 \cdot 10^6$ Πa; $E \cong 2 \cdot 10^{11}$ Πa, and inversely proportional to δ/D ($\delta \approx 5–10$ mm; $D \approx 300–100$ mm), the dependence under study could be written as

$$\Delta S = D^2 \cdot \frac{P}{E} \cdot \frac{1}{\delta/D} \cdot \tilde{f}_0 = \tilde{f}_0 \cdot \frac{D^3 \cdot P}{\delta \cdot E}$$

where $\tilde{f}_0$ is a certain constant. Hence, in the dependence under study there is only one constant left to be obtained. Theoretical investigation shows that the constant $\tilde{f}_0$ is equal to $\pi/4 \cong 0.785$.

Exercise 2. It is known that the laminar flow of a viscous incompressible fluid in a circular pipe loses stability and becomes turbulent. It is required to investigate the dependence of the critical flow velocity v_{cr} at which this transition happens, taking the critical velocity as a function of three parameters: pipeline diameter d, dynamic viscosity μ and density ρ of the fluid. In other words it is required to investigate the dependence $v_{cr} = f(d, \mu, \rho)$.

Solution. In the given case $A = v_{cr}, a_1 = d; a_2 = \mu; a_3 = \rho; n = 3$. The dimensions of the parameters in the SI system are: $[v_{cr}] = \text{m s}^{-1}$; $[d] = \text{m}$; $[\mu] = \text{kg m}^{-1}\,\text{s}^{-1}$; $[\rho] = \text{kg m}^{-3}$. From this it follows that all three arguments of the function are dimensionally-independent, i.e. $k = n = 3$, and the number of arguments in the dimensionless writing of the function under study may be reduced by three, i.e. to 0.

The single dimensionless complex Π can be written as $\Pi = v_{cr} \cdot d/(\mu/\rho)$. The sought dependence then takes the especially simple form

$$\Pi_{cr} = \frac{v_{cr} \cdot d}{(\mu/\rho)} = \text{const.} \tag{6.17}$$

The ratio μ/ρ is called the *kinematic viscosity* of the fluid and is denoted by ν ($[\nu] = L^2/T$). In the SI system the unit of kinematic viscosity is stokes (St); 1 St $= 10^{-4}$ m^2 s^{-1}; the viscosity of water is ≈ 0.01 St $= 1$ cSt (centistokes) $= 10^{-6}$ m^2 s^{-1}.

The dimensionless parameter Π_{cr} determining the transition of laminar flow to turbulent flow is called the *critical Reynolds number* and is denoted by Re_{cr}. The theory and experiments have shown that $Re_{cr} \cong 2300$. At $Re < Re_{cr}$ the flow is laminar whereas at $Re > Re_{cr}$ it is turbulent.

Exercise 3. A ball of mass m and diameter D is dropped in a viscous fluid with density ρ and viscosity ν under the action of gravity (gravity acceleration – g) with constant velocity v. The dependence of this velocity on the governing parameters: $v = f(m, D, \rho, \nu, g)$ is to be investigated. Using the Π-theorem it is required to write the sought dependence in dimensional form.

Solution. Dimensions of the parameters ($n = 5$), in this problem ($A = v$, $a_1 = m$, $a_2 = D$, $a_3 = \rho$, $a_4 = \nu$, $a_5 = g$), in the SI system are $[v] = $ m s^{-1}; $[m] = $ kg; $[D] = $ m; $[\rho] = $ kg m^{-3}; $[\nu] = $ m^2 s^{-1}; $[g] = $ m s^{-2}. The maximum number of dimensionally-independent parameters among the arguments is equal to three (as such parameters can be taken for example ρ, D and g), the number of arguments in the sought dimensionless writing may be reduced to two ($5 - 3 = 2$).

Setting up dimensionless complexes Π, Π_1 and Π_2

$$\Pi = \frac{v}{g^{1/2} D^{1/2}}, \quad \Pi_1 = \frac{m}{\rho \cdot D^3}, \quad \Pi_2 = \frac{\nu}{g^{1/2} D^{3/2}},$$

the sought dependence can be represented as follows

$$\Pi = f(\Pi_1, \Pi_2)$$

or

$$v = \sqrt{gD} \cdot f\left(m/\rho D^3, \nu/\sqrt{gD^3}\right).$$

It is seen that the dependence contains in fact not five dimensional arguments, but only two dimensionless complexes.

Exercises.

1. Using the Π-theorem, it is required to write in dimensional form the dependence of the resistance force F experienced by a submarine

moving in water (density ρ, kinematic viscosity ν) with velocity v, if we accept that this force depends also on the diameter of the submarine cross-section D and the submarine length L, that is $F = f(v, \rho, \nu, D, L)$.

Answer. $F/(\rho v^2 D^2) = \tilde{f}(vD/\nu, D/L)$.

2. Using the Π-theorem, it is required to write in dimensional form the dependence of the oscillation period T of a mathematical pendulum (massive point on non-stretchable line), if the mass of the latter is m, the length of the line is L and the acceleration due to gravity is g, that is $T = f(m, L, g)$.

   Answer. $T/\sqrt{L/g} = \text{const.}$

3. Using the Π-theorem, it is required to write in dimensional form the dependence of the time T of complete outflow of fluid (density ρ and viscosity ν) from a tank car with diameter D and length L. The outflow happens under gravity (acceleration due to gravity g) through the drain system with flow area s located at the tank bottom. The sought function is $T = f(\rho, \nu, g, D, L, s)$.

   Answer. $T/\sqrt{D/g} = \tilde{f}(\sqrt{g}D^{3/2}/\nu, L/D, s/D^2)$.

4. Using the Π-theorem, it is required to write in dimensional form the same dependence as in the previous exercise with the single distinction that the outflow happens not only under the action of gravity but also under the action of positive pressure Δp created inside the tank. The sought function is $T = f(\Delta p, \rho, \nu, g, D, L, s)$.

   Answer. $T/\sqrt{D/g} = \tilde{f}(\sqrt{g}D^{3/2}/\nu, \Delta p/(\rho g D), L/D, s/D^2)$.

5. Using the Π-theorem, it is required to write in dimensional form the dependence of the volumetric flow rate Q of a fluid with density ρ and viscosity ν in an inclined pipeline (inclination angle α) having cross-section area S_0, if this flow represents voluntary (gravity) flow.

   The flow happens under the action of the gravity force projection $g \sin\alpha$, the area of the pipeline cross-section filled with fluid is $S < S_0$. The sought function is $Q = f(g \sin\alpha, S, S_0, \rho, \nu)$.

   Answer. $Q/[S_0^{5/4} \cdot (g \sin\alpha)^{1/2}] = \tilde{f}(S/S_0, (g \sin\alpha)^{1/2} \cdot S_0^{3/4}/\nu)$.

7
Physical Modeling of Phenomena

The main advantage of the dimensional theory is that it opens up possibilities to use the similarity laws of physical phenomena and allows modeling of these phenomena through replacing them in nature by similar phenomena on a reduced or enlarged scale under experimental conditions.

7.1
Similarity of Phenomena and the Principle of Modeling

In order to elucidate the essence of modeling, let us consider several examples.

Assume that it is required to calculate the radius of a circle inscribed in a triangle whose sides are very large, for example 1, 2 and 3 km. This problem may be easily solved by simple algebraic calculation. However, if it was necessary to perform this measurement experimentally, one could proceed as follows: on a sheet of paper draw a triangle on a reduced scale similar to the given triangle, for example a triangle with sides 10, 20 and 30 cm. The similarity factor of this triangle to the full-scale one would be equal to 10 000. Inscribing a circle in the depicted triangle, it is easy to measure its radius. Then the obtained number multiplied by the similarity factor, i.e. 10 000, yields the sought radius of the circle inscribed in the full-scale triangle. Of course in the case under consideration we are dealing with a simple geometric similarity and geometric modeling but it clarifies the essence of modeling in the general case.

Let it be required to determine experimentally the dump time of a tank of complex geometrical form and very large size. In order to solve this problem it is decided to make a copy of the tank on a reduced scale, fill it with some model liquid and then measure the dump time. The question arises as to whether it is enough to provide merely geometric similarity of full-scale and model tanks and to use in experiments the same liquid or is it necessary to replace the liquid with another one with specially selected properties. Moreover, is it necessary to determine the ratio between the measured dump time and that actually taking place in the full-scale object. The answers to these questions should give the theory of simulation (modeling).

Finally, one more example from engineering practice. It is required to clarify whether the river dam will withstand the dynamic head of flooding water.

Modeling of Oil Product and Gas Pipeline Transportation. Michael V. Lurie

ISBN: 978-3-527-40833-7

For this purpose a reduced size copy of the dam is made, mounted in an experimental channel in a hydrological laboratory. It is evident that if the dam were made from the same material as in the natural conditions, i.e. from ferroconcrete, the dam would withstand any water head. It is necessary to determine the velocity of water in the experimental channel in order to model the river head. When reducing the linear sizes of the considered phenomenon, it is insufficient to provide geometric similarity. It is necessary especially to change the scales of many other parameters of the phenomenon.

Similar problems take place in different fields of engineering: hydraulic engineering, aviation, transport, the storage of oil and gas and so on.

Definition. Two phenomena are called *similar* when, from the given parameters of one phenomenon, analogous parameters of another phenomenon are determined by simple recalculation of the same kind as for transition from one system of measurement units to another. Each of such phenomena is called a *model* of another phenomenon from this set of phenomena.

7.2 Similarity Criteria

Let us determine the necessary and sufficient conditions of two phenomena to be similar. Such conditions are called *similarity criteria.*

Let a phenomenon be such that a certain physical quantity A is determined by a set of physical parameters $a_1, a_2, \ldots, a_n$, so that

$$A = f(a_1, a_2, \ldots, a_n). \tag{7.1}$$

The model under consideration consists of the dependence of an analogous physical quantity A' on the same physical parameters the numerical values of which $a_1', a_2', \ldots, a_n'$ differ from those determining the quantity A. Thus, we have

$$A' = f\left(a_1', a_2', \ldots, a_n'\right). \tag{7.2}$$

In accordance with the I-theorem both dependences (7.1) and (7.2) may be rewritten in dimensionless form as follows

$$\begin{aligned} \Pi &= f\left(\Pi_1, \Pi_2, \ldots, \Pi_{n-k}\right), \\ \Pi' &= f\left(\Pi_1', \Pi_2', \ldots, \Pi_{n-k}'\right), \end{aligned} \tag{7.3}$$

where k is the number of dimensionally-independent parameters among the quantities $a_1, a_2, \ldots, a_n$.

Relations (7.3) show that if the parameters $a_1', a_2', \ldots, a_n'$ are chosen such that the following conditions are obeyed

$$\Pi_1' = \Pi_1, \quad \Pi_2' = \Pi_2, \ldots, \quad \Pi_n' = \Pi_n. \tag{7.4}$$

Then the following condition would also be satisfied

$$\Pi' = \Pi. \tag{7.5}$$

The value of the parameter A could be found by simple recalculation of the parameter A' through

$$A = A' \cdot \frac{a_1^{m_1} a_2^{m_2} \cdots a_k^{m_k}}{a_1'^{m_1} a_2'^{m_2} \cdots a_k'^{m_k}} \tag{7.6}$$

and the considered phenomena would, by definition, be similar.

Thus, necessary and sufficient conditions of two phenomena to be similar are equalities of the dimensionless complexes determining these phenomena, namely conditions (7.4). Hence, the dimensionless parameters $\Pi_1, \Pi_2, \ldots, \Pi_{n-k}$ are the sought similarity criteria.

7.3 Modeling of Viscous Fluid Flow in a Pipe

As an example of two similar phenomena, consider the modeling of the stationary flow of a viscous incompressible fluid in a model pipe with reduced size as compared to the full-scale one.

In accordance with the results given in Section 1.6 this dependence is

$$\frac{\Delta p}{\frac{L}{d} \cdot \frac{\rho v^2}{2}} = \lambda(Re, \varepsilon).$$

Denote through $\rho', v', d', \mu', \Delta', L', \Delta p'$ the values of the hydrodynamic parameters relating to the flow in the model pipe. The same parameters without the superscript "prime" refer to the phenomena to be modeled. Then

$$\frac{\Delta p'}{\frac{L'}{d'} \cdot \frac{\rho' v'^2}{2}} = \lambda(Re', \varepsilon').$$

The parameters of the model pipe and the flow regime in it are chosen to obey the relations

$$Re = Re', \ \ \varepsilon = \varepsilon' \tag{7.7}$$

or

$$\frac{vd}{\nu} = \frac{v'd'}{\nu'}, \quad \frac{\Delta}{d} = \frac{\Delta'}{d'}.$$

In so doing we ensure the equality

$$\frac{\Delta p'}{\frac{L'}{d'} \cdot \frac{\rho' v'^2}{2}} = \frac{\Delta p}{\frac{L}{d} \cdot \frac{\rho v^2}{2}}$$

or

$$\Delta p = \Delta p' \cdot \frac{L}{L'} \cdot \frac{d'}{d} \cdot \frac{\rho}{\rho'} \cdot \left(\frac{\nu}{\nu'}\right)^2. \tag{7.8}$$

If we now take the ratio d'/d, showing how many times smaller are the sizes of the model pipe than the sizes of the full-scale pipe (factor of geometric similarity), and the ratio ν'/ν of the viscosities of fluids flowing in full-scale and model pipes, respectively, we could determine the fluid flow velocity and roughness of the pipe walls of the experimental (model) installation needed to achieve the similarity

$$v' = v \cdot \frac{d}{d'} \cdot \frac{\nu'}{\nu}, \; \Delta' = \Delta \cdot \frac{d'}{d}. \tag{7.9}$$

Formulas (7.9) state that the similarity in the case under consideration is afforded by fulfilling two conditions: equality of the Reynolds numbers and equality of the wall relative roughness. Consequently, the Reynolds number and relative roughness serve as similarity criteria for the problem on the flow of a viscous fluid in a pipe.

7.4 Modeling Gravity Fluid Flow

This type of flow has already been considered in Section 3.7. We now set a question on the physical modeling of this process. Let the flow rate Q of the fluid in a pipe inclined to the horizontal at an angle α be given. It is required to determine the depth h of the fluid filling the pipe cross-section, that is, the function

$$h = f(Q, d, g \sin\alpha, \nu, \Delta).$$

Rewrite the dependence to be sought in dimensionless form using the Π-theorem. Among five arguments of this function there are two dimensionally-independent, for example ν, d, therefore in dimensionless form the number of independent arguments will be reduced from five to three.

$$\frac{h}{d} = \tilde{f}\left(\frac{(Q/d^2) \cdot d}{\nu}, \frac{\sqrt{gd \sin\alpha} \cdot d}{\nu}, \frac{\Delta}{d}\right).$$

It is more convenient in this dependence to use the ratio of the first argument to the second one instead of the second argument. As a result we have

$$\frac{h}{d} = \tilde{f}_1\left(\frac{(Q/d^2) \cdot d}{\nu}, \frac{(Q/d^2)}{\sqrt{gd \sin\alpha}}, \frac{\Delta}{d}\right).$$

In the first argument it is easy to recognize the Reynolds number Re of the flow calculated from the velocity $v = Q/d^2$ and the kinematic viscosity $\nu = \mu/\rho$, so that the first similarity criteria would be $\Pi_1 = Re$. The second argument is called the Froude number, *Fr*. The Froude number is, in general, equal to v^2/gd, thus in our case we are dealing with a Froude number calculated from the velocity Q/d^2 and the component of the acceleration due togravity $g\sin\alpha$. Nevertheless, the second similarity criterion can be taken as $\Pi_2 = Fr$. Finally, the third argument of this dependence $\Pi_3 = \varepsilon$ is the relative roughness of the pipe's internal wall surface. Hence, the dependence under study is

$$\frac{h}{d} = \hat{f}(Re, Fr, \varepsilon). \tag{7.10}$$

To model this phenomenon it is necessary to provide the following similarity conditions

$$Re' = Re \;\Rightarrow\; \frac{Q'}{\nu'd'} = \frac{Q}{\nu d} \;\Rightarrow\; Q' = Q \cdot \left(\frac{\nu'}{\nu}\right) \cdot \left(\frac{d'}{d}\right),$$

$$Fr' = Fr \;\Rightarrow\; \frac{Q'^2}{g\sin\alpha' \cdot d'^5} = \frac{Q^2}{g\sin\alpha \cdot d^5},$$

$$\Rightarrow\; Q' = Q \cdot \sqrt{\frac{\sin\alpha'}{\sin\alpha}} \cdot \left(\frac{d'}{d}\right)^{5/2}, \tag{7.11}$$

$$\varepsilon' = \varepsilon \;\Rightarrow\; \frac{\Delta'}{d'} = \frac{\Delta}{d} \;\Rightarrow\; \Delta' = \Delta \cdot \frac{d'}{d}.$$

If these conditions are obeyed, the depth h to which the pipe is filled can be calculated using the formula

$$\frac{h'}{d'} = \frac{h}{d} \;\Rightarrow\; h = h' \cdot \frac{d}{d'}. \tag{7.12}$$

In order for the first two conditions to be consistent, the following equalities should be true

$$\left(\frac{\nu'}{\nu}\right) \cdot \left(\frac{d'}{d}\right) = \sqrt{\frac{\sin\alpha'}{\sin\alpha}} \cdot \left(\frac{d'}{d}\right)^{5/2}$$

or

$$\sin\alpha' = \sin\alpha \cdot \left(\frac{\nu'}{\nu}\right)^2 \cdot \left(\frac{d}{d'}\right)^3. \tag{7.13}$$

If in the model the same liquid as in the full-scale pipe is used, then a different slope of the pipe should be used in the model, it must be chosen in accordance with relation (7.13). If $\alpha' = \alpha$, then the viscosity of the fluid used in the model should be different, namely $\nu' = \nu \cdot (d'/d)^{3/2}$.

For example, let the linear size of the experimental pipe be 1/5th that of the full-scale pipe, that is, $d'/d = 1/5$, and the viscosity ν' of the model liquid be five times less than the viscosity of the natural liquid. Then the conditions for the experiment should be: $Q' = 1/5 \cdot 1/5 \cdot Q = 0.04 \cdot Q$; $\sin\alpha' = 1/25 \cdot 5^3 \cdot \sin\alpha = 5 \cdot \sin\alpha$; $\Delta' = 0.2 \cdot \Delta$, that is, the fluid flow rate in the experiment has to be reduced by a factor of 25, the internal surface of the pipe should be so polished that the absolute roughness is reduced by five times as compared to the full-scale pipe and the sine of the inclination angle of the model pipe should be increased by five times. Thus, the depth h of the flow in the full-scale pipe has to be five times greater than the model depth h', that is, $h = 5 \cdot h'$.

In experiments it is not always possible to obey all the required similarity conditions, therefore only the main conditions are satisfied. If, for example, in a given case a flow of diesel fuel with viscosity 3 cSt is to be investigated, then benzene with viscosity $\cong$0.6 cSt could be taken for the model. Reduction of the flow rate and enhancement of the pipe slope (at small values of α) present no special problems. It is far more complicated to fulfill the last similarity condition and so sometimes it is neglected.

7.5 Modeling the Fluid Outflow from a Tank

Let there be a railway tank with a boiler diameter D and length L or, in general, a tank of arbitrary geometric form intended to transport oil with density ρ and kinematic viscosity ν. The tank is provided with bottom drain equipment with the area of the flowing cross-section S. It is required to design an experimental installation to model the process and to investigate the dependence of the time T of the oil outflow on the parameters of the liquid and tank.

The dependence under study $T = f(D, L, \rho, \nu, g, S)$ in dimensionless form is written as

$$\frac{T}{\sqrt{D/g}} = \tilde{f}\left(\frac{L}{D}, \frac{S}{D^2}, \frac{\sqrt{gD} \cdot D}{\nu}\right).$$

Therefore the similarity criteria are:

$$\Pi_1 = \frac{L}{D}, \; \Pi_2 = \frac{S}{D^2} \quad \text{the criteria of geometric similarity;}$$

$$\Pi_3 = \frac{g^{1/2} \cdot D^{3/2}}{\nu} \quad \text{the criterion of dynamic similarity.}$$

At the fulfillment of the first two criteria at which the model tank should be similar to the full-scale one, the phenomena would be similar under the condition

$$\frac{g^{1/2} \cdot D'^{3/2}}{\nu'} = \frac{g^{1/2} \cdot D^{3/2}}{\nu},$$

where the superscripts "prime" refer to the model parameters. The last relation shows that the sufficient condition for the dynamic similarity of the model tank to the full-scale one is the condition

$$\nu' = \nu \cdot \left(\frac{D'}{D}\right)^{3/2}.$$

If this condition is fulfilled, then $T'/\sqrt{D'/g} = T/\sqrt{D/g}$ or

$$T = T' \cdot \sqrt{\frac{D}{D'}}. \tag{7.14}$$

From this follows, in particular, that for modeling of oil outflow on the installation with sizes distinct from the full-scale tank, it is necessary to use a liquid with velocity also differing from the oil under study. Let the experimental tank be a cylindrical tank 10 times smaller than the full-scale one, that is $D/D' = 10$. Then $\nu' \cong 0.0316 \cdot \nu$, i.e. the kinematic viscosity of the liquid used in the experiments, should be 32 times lower than the viscosity of the oil being modeled. If this oil has sufficiently high viscosity then it is easy to select such a liquid, otherwise the problem is complicated. The time T of oil outflow is determined by the time T' measured in the experiment. In accordance with the formula (7.14) it is $T \cong 3.162 \cdot T'$.

7.6 Similarity Criteria for the Operation of Centrifugal Pumps

As already described in Section 4.2.1 pumps are equipment to make liquids flow against a pressure force, that is, in the direction from lesser pressure p_{suc} at the suction line to the greater pressure p_d at the line of discharge under pressure. Of course, it is possible only by the work of external energy sources (mechanical, thermal, electrical and so on).

Centrifugal pumps represent a variety of pump in which the centrifugal force, acting on fluid particles rotating in the impeller, makes the liquid flow against the pressure. The propulsive device of the impeller or, as it is called, the *pump drive*, can be an internal combustion engine, vapor-gas turbine or any other source of rotational moment (see Fig. 4.1).

The volumetric fluid flow rate Q (or, as it is called, the *feed*) depends on the pressure drop $\Delta p = p_d - p_{suc}$ which the liquid should overcome at its flow from the impeller center to the periphery. The greater the pressure drop the lesser is the flow rate of the liquid.

If the area σ of the outlet branch pipe at the discharge line is given by the pump design, the fluid velocity Q/σ is a function of the impeller diameter D_{im} (centrifugal pumps may have accessory impellers), the angular velocity ω of the impeller rotation (centrifugal pumps may be supplied with equipment to change ω), the density ρ and viscosity ν of the fluid to be pumped, so that

$$\frac{Q}{\sigma} = f(\Delta p, D_{im}, \omega, \rho, \nu). \tag{7.15}$$

The dependence (7.15) is called the $(Q - p)$ characteristic of the centrifugal pump and is to a large extent determined by its constructive peculiarities.

Among the arguments of the function (7.15) there are three dimensionally-independent parameters, for example D_{im}, ω, ρ, therefore its dimensionless form could be reduced to

$$\frac{Q/\sigma}{\omega D_{im}} = \tilde{f}\left(\frac{\Delta p}{\rho\omega^2 D_{im}{}^2}, \frac{\nu}{\omega D_{im} \cdot D_{im}}\right).$$

The last argument represents none other than a quantity inversely proportional to the Reynolds number *Re*. This parameter reflects the effect of viscosity on the characteristic of the pump operation.

If we introduce into consideration the quantity $\Delta H = \Delta p/\rho g$ (the *differential head of the pump*) and solve Eq. (7.15) with respect to this head, we get the so-called $(Q - \Delta H)$ characteristic of the pump

$$\Delta H = \omega^2 D_{im}{}^2 \cdot F\left(\frac{Q}{\omega D_{im}}, \frac{\omega D_{im}{}^2}{\nu}\right). \tag{7.16}$$

Here the constants σ and g are taken into account by the form of the function F.

It is interesting to note that the density ρ of the liquid to be pumped does not enter into the dependence (7.16), i.e. the form of the latter is true for the pump operated by liquid of any density. As for the influence of the viscosity ν on the form of $(Q - \Delta H)$ characteristic of centrifugal pumps, it is small. Reynolds numbers $\omega D_{im}{}^2/\nu$ have rather large values ($\approx 10^6$) due to the high rotation velocity of the pump impeller ($\omega \approx 300\ \text{s}^{-1}$, $\omega = 2\pi \cdot n$; $n \approx 3000$ rpm; $D_{im} \approx 0.2$–0.7 m; $\nu \approx 1$–$10 \cdot 10^{-6}\ \text{m}^2\text{s}^{-1}$), therefore in practice their variation only slightly affects the $(Q - \Delta H)$ characteristic of the centrifugal pump and thus the influence of the pumping fluid viscosity is also small. The latter conclusion is of course true only up to certain limits.

If we ignore the influence of pumping fluid viscosity on the $(Q - \Delta H)$ characteristic of the centrifugal pump, then it can be represented in the simple form

$$\Delta H = \omega^2 D_{im}{}^2 \cdot F\left(\frac{Q}{\omega D_{\kappa}}\right). \tag{7.17}$$

From the derived formulas some practically important conclusions follow:

- If a centrifugal pump operating with angular velocity ω_0 or revolutions per minute n_0 has a characteristic $\Delta H = F_*(Q)$, the same pump working with varied rotation frequency ω or revolutions per minute n, has the characteristic

$$\Delta H = \left(\frac{\omega}{\omega_0}\right)^2 \cdot F_*\left(\frac{\omega_0}{\omega} \cdot Q\right); \tag{7.18}$$

- If a centrifugal pump, operating with an impeller of diameter $D_{\kappa 0}$ has the characteristic $\Delta H = F_*(Q)$, the same pump operating with another impeller of diameter D_{im} has the characteristic

$$\Delta H = \left(\frac{D_{im}}{D_{im0}}\right)^2 \cdot F_* \left(\frac{D_{im0}}{D_{im}} \cdot Q\right); \tag{7.19}$$

- If a centrifugal pump, operating with rotational velocity ω_0 and impeller diameter D_{im0} has the characteristic $\Delta H = F_*(Q)$, the same pump operating with varied rotational frequency ω and impeller diameter D_{im} has the characteristic

$$\Delta H = \left(\frac{\omega D_{im}}{\omega_0 D_{im0}}\right)^2 \cdot F_* \left(\frac{\omega_0 D_{im0}}{\omega D_{im}} \cdot Q\right). \tag{7.20}$$

Rules (7.18)–(7.20) allow us to change the $(Q - \Delta H)$ characteristic of centrifugal pumps by changing the rotation speed or/and the impeller diameter.

In many cases, as has been said, the $(Q - \Delta H)$ characteristics of centrifugal pumps are represented in the form of a parabola

$$\Delta H = F_*(Q) = a - b \cdot Q^2 \tag{7.21}$$

where a and b are approximation factors. If we now change the impeller diameter from D_{im0} to D_{im} and the rotation frequency from ω_0 to ω, the $(Q - \Delta H)$ characteristic of the same pump takes the form

$$\Delta H = a \cdot \frac{\omega^2 {D_{im}}^2}{\omega_0^2 D_{im0}^2} - b \cdot Q^2, \tag{7.22}$$

that is the parabola graphic in the plane $(Q, \Delta H)$ undergoes a displacement along the H-axis by

$$a \cdot \left(1 - \frac{\omega^2 {D_{im}}^2}{\omega_0^2 {D_{im0}}^2}\right).$$

Exercise. The diameter of a centrifugal pump impeller is 490 mm and the impeller velocity is 3200 rpm. The pump has the following characteristic

$$\Delta H = 331 - 0.451 \cdot 10^{-4} \cdot Q^2$$

(ΔH in m, Q in m^3 h^{-1}). It is required to determine the characteristic of the same pump if we reduce the impeller diameter to 480 mm and the number of revolutions per minute to 3000.

Solution. In accordance with Eq. (7.22) the sought characteristic has the following form

$$\Delta H = 331 \cdot \left(\frac{3000 \cdot 480}{3200 \cdot 490}\right)^2 - 0.451 \cdot 10^{-4} \cdot Q^2$$
$$= 279 - 0.451 \cdot 10^{-4} \cdot Q^2.$$

Exercises.

1. How does the $(Q - \Delta H)$ characteristic of a centrifugal pump vary if the frequency speed of the impeller is increased from 3000 to 3200 rpm? The given characteristic of the pump is: $\Delta H = 360 - 0.42 \cdot 10^{-4} \cdot Q^2$, where ΔH is measured in m, Q in $m^3\ h^{-1}$.

   Answer. $\Delta H = 410 - 0.42 \cdot 10^{-4} \cdot Q^2$.

2. How does the $(Q - \Delta H)$ characteristic of a centrifugal pump vary if the frequency speed of the impeller is increased from 3000 to 3200 rpm? The given characteristic of the pump is: $\Delta H = 360 - 0.42 \cdot 10^{-4} \cdot Q^{1.75}$, where ΔH is measured in m, Q in $m^3\ h^{-1}$.

   Answer. $\Delta H = 410 - 0.375 \cdot 10^{-4} \cdot Q^{1.75}$.

3. How does the $(Q - \Delta H)$ characteristic of a centrifugal pump vary if the impeller diameter is reduced from 480 to 470 mm. The given characteristic of the pump is: $\Delta H = 360 - 0.42 \cdot 10^{-4} \cdot Q^{1.75}$, where ΔH is measured in m, Q in $m^3\ h^{-1}$.

   Answer. $\Delta H = 345 - 0.436 \cdot 10^{-4} \cdot Q^{1.75}$.

8
Dimensionality and Similarity in Mathematical Modeling of Processes

In the previous chapter we have seen that dimensionless similarity criteria $\Pi_1, \Pi_2, \ldots, \Pi_{n-k}$ appear in the considered problems after fixation of a set of governing parameters $a_1, a_2, \ldots, a_n$ by way of heuristic reasoning. Hence, it is by no means necessary to know which equations satisfy these parameters and with which physical laws they are connected. It is enough to know only the dimensions of these parameters in order to set up dimensionless similarity criteria by which the class of considered phenomena is characterized.

8.1
Origination of Similarity Criteria in the Equations of a Mathematical Model

A somewhat different approach takes place in the mathematical modeling of phenomena in construction systems of algebraic or differential equations with initial and boundary conditions in which we would like to see an adequate model of the considered phenomena or process. Dimensionless similarity parameters originate in these models in a strictly specified way of bringing model equations to a dimensionless form.

Let us illustrate the afore said by a simple example. It is known that one-dimensional oscillations of a pont weight with mass m on an elastic spring around an equilibrium position ($x = 0$) are described by an ordinary differential equation

$$m\frac{\mathrm{d}^2 x}{\mathrm{d}t^2} = -k \cdot x \tag{8.1}$$

where $x(t)$ is the linear dependence of the weight coordinate on time and k is the restoring force factor. The initial, at $t = 0$, position and velocity $\dot{x}(0) = v_0$ of the weight are also given, that is the mathematical model of the process.

Let us bring the model equation to dimensionless form. First we introduce dimensionless variables

$$\bar{x} = \frac{x}{L} \quad \text{and} \quad \bar{\dot{x}} = \frac{\dot{x}}{v_0}, \qquad (L \neq 0, v_0 \neq 0).$$

Modeling of Oil Product and Gas Pipeline Transportation. Michael V. Lurie

ISBN: 978-3-527-40833-7

Then the dimensionless time $\bar{t} = t/(L/v_0)$ is determined. In new variables the differential equation transforms to

$$m \cdot \frac{L \cdot \mathrm{d}^2\bar{x}}{(L/v_0)^2 \cdot \mathrm{d}\bar{t}^2} = -k \cdot (L \cdot \bar{x})$$

or

$$\frac{\mathrm{d}^2\bar{x}}{\mathrm{d}\bar{t}^2} = -\frac{kL^2}{mv_0^2} \cdot \bar{x}. \tag{8.2}$$

The initial conditions take the form $\bar{x}(0) = 1$; $\dot{\bar{x}} = 1$.

Hence, the oscillation of different weights with different mass on springs with different elasticity caused by different initial conditions is in fact described by a mathematical model containing only *one dimensionless parameter* $\Pi = kL^2/mv_0^2$. If in two situations this parameter appears to be identical, the solutions of Eq. (8.2) are indistinguishable and consequently we have to deal with *similar* situations.

It is easy to verify that the solution of the Eq. (8.2) is

$$\bar{x}(\bar{t}) = \cos\left(\sqrt{\Pi}\bar{t}\right) + \frac{1}{\sqrt{\Pi}} \cdot \sin\left(\sqrt{\Pi}\bar{t}\right).$$

Returning to dimensional quantities, we obtain

$$x(t) = L \cdot \cos\left(\sqrt{\frac{k}{m}} \cdot t\right) + \sqrt{\frac{m}{k}} \cdot v_0 \cdot \cos\left(\sqrt{\frac{k}{m}} \cdot t\right).$$

As is known $\sqrt{k/m} = \omega$ is the frequency of harmonic oscillations of the weight and the amplitude h is the square root of the sum of the squared factors of the sine and cosine

$$h = \sqrt{L^2 + \frac{mv_0^2}{k}} = L \cdot \sqrt{1 + \frac{1}{\Pi}}.$$

8.2
One-Dimensional Non-Stationary Flow of a Slightly Compressible Fluid in a Pipeline

The theory of one-dimensional non-stationary flows of a slightly compressible fluid in a pipeline was considered in detail in Sections 5.1–5.3. The main equations modeling such flows are

$$\begin{cases} \dfrac{\partial p}{\partial t} + c^2 \dfrac{\partial(\rho v)}{\partial x} = 0 \\ \rho\left(\dfrac{\partial v}{\partial t} + v\dfrac{\partial v}{\partial x}\right) = -\dfrac{\partial p}{\partial x} - \dfrac{\lambda(Re, \varepsilon)}{d} \cdot \dfrac{\rho v^2}{2} \end{cases} \tag{8.3}$$

in which the quantity

$$c = \frac{1}{\sqrt{\dfrac{\rho_0}{K} + \dfrac{\rho_0 d}{\delta \cdot E}}} \tag{8.4}$$

is called the velocity of the pressure wave propagations in pipeline.

Consider a certain problem on the calculation of non-stationary flow of a slightly compressible fluid in pipeline taking Eqs. (8.3) as the mathematical model of this flow.

Problem. Let there be at the pipeline section $0 \leq x \leq L$ stationary fluid flow with velocity v_0. However, starting from some time a valve located at the end cross-section of the pipeline $x = L$ begins to vary its opening level with frequency ω. It is required to reveal the similarity criteria of this phenomenon.

Solution. Introduce the following dimensionless variables marked by the horizontal bar at the top

$$t = \frac{1}{\omega} \cdot \bar{t}, \quad x = L \cdot \bar{x}, \quad v = v_0 \cdot \bar{v}, \quad \rho = \rho_0 \cdot \bar{\rho}, \quad p = p_0 \cdot \bar{p}.$$

Equations (8.3) in the new variables take the form

$$\begin{cases} \dfrac{p_0}{1/\omega} \cdot \dfrac{\partial \bar{p}}{\partial \bar{t}} + c^2 \cdot \dfrac{\rho_0 v_0}{L} \cdot \dfrac{\partial(\bar{\rho}\bar{v})}{\partial \bar{x}} = 0 \\ \rho_0 \bar{\rho} \left(\dfrac{v_0}{1/\omega} \cdot \dfrac{\partial \bar{v}}{\partial \bar{t}} + \dfrac{v_0^2}{L} \cdot \bar{v} \dfrac{\partial v}{\partial \bar{x}} \right) = -\dfrac{p_0}{L} \cdot \dfrac{\partial \bar{p}}{\partial \bar{x}} - \dfrac{\lambda(Re, \varepsilon)}{d} \cdot \rho_0 v_0^2 \cdot \dfrac{\overline{\rho v}^2}{2} \end{cases}$$

If now we take $p_0 = \rho_0 v_0 c$, the system of equations simplifies to

$$\begin{cases} \dfrac{\omega L}{c} \cdot \dfrac{\partial \bar{p}}{\partial \bar{t}} + \dfrac{\partial(\bar{\rho}\bar{v})}{\partial \bar{x}} = 0 \\ \dfrac{\omega L}{c} \cdot \dfrac{\partial \bar{v}}{\partial \bar{t}} + \dfrac{v_0}{c} \cdot \bar{v} \dfrac{\partial \bar{v}}{\partial \bar{x}} = -\dfrac{\partial \bar{p}}{\partial \bar{x}} - \dfrac{\lambda(Re, \varepsilon) \cdot L}{d} \cdot \dfrac{v_0}{c} \cdot \dfrac{\overline{\rho v}^2}{2} \end{cases} \tag{8.5}$$

From Eqs. (8.5) it is seen that there are three dimensionless criteria governing the class of problems under consideration and differing from each other only by the numerical values of the parameters entering in these equations

$$\Pi_1 = \frac{v_0}{c}; \quad \Pi_2 = \frac{\omega L}{c}; \quad \Pi_3 = \frac{\lambda L}{d}.$$

The first of these criteria is called the *Mach number*. The ratio of the first criterion to the second one $\Pi_1/\Pi_2 = v_0/(\omega L)$ is called the *Strouchal number* $St = v_0/(\omega L)$.

The Mach number of the fluid or gas flow in a main pipeline is, as a rule, very small (for example, for fluid flow $M \approx 0.001$; for gas flow $M \approx 0.03$), therefore it often turns out that the second term in the left-hand side of the last equation (8.5) can be neglected in comparison with the first one. If the flow velocity v_0 is comparable to the sound velocity c, as it may be, for example, at the gas outflow from a pipeline through a short nozzle, the second term would become almost the primary term and could not be ignored.

The Strouchal number characterizes the degree of process non-stationarity. If this number is great ($St = v_0/\omega L \gg 1$, that is, the characteristic time of the process $1/\omega$ is large compared to L/v_0 the time of fluid particle passage through the pipeline), the non-stationarity degree is small and the process is close to stationary. If the Strouchal number is small (that is the frequency of the process ω is large compared to v_0/L) then the non-stationarity of the process cannot be neglected.

The third criterion $\lambda L/d$ does not have a special name. It characterizes the magnitude of the resistance to the fluid friction in the pipeline.

8.3
Gravity Fluid Flow in a Pipeline

Consider now problems connected with thegravity flow of incompressible fluid in a pipeline (see Sections 3.7 and 7.3). Earlier we were dealing with stationary gravity fluid flow, that is with flow in which all the hydrodynamic parameters at each pipeline cross-section remained constant, now we will consider the general case of non-stationary gravity flow in a descending pipeline section characterized by the slope angle to the horizontal α ($\alpha < 0$). The differential equations describing such flow are

$$\begin{cases} \dfrac{\partial \rho S}{\partial t} + \dfrac{\partial \rho v S}{\partial x} = 0 \\[2ex] \dfrac{\partial \rho v S}{\partial t} + \dfrac{\partial}{\partial x}\left(\rho v^2 S + \rho g \cos\alpha \cdot \displaystyle\int_0^S S\, dh\right) \\[2ex] \qquad = -\rho g S \sin\alpha - \dfrac{\rho g S \cos\alpha}{C_{Sh}{}^2 R_h} \cdot v|v| \end{cases}$$

Here $S(x,t)$ is the area of the pipeline cross-section filled by fluid; $R_h(S)$ the hydraulic radius of the flow; $h(S)$ the depth of fluid in the pipeline cross-section; C_{Sh} the Chezy factor (see Section 3.7).

Let v_0 be the fluid velocity at the inlet to the descending section of the pipeline and $\rho \cong \rho_0 = \text{const}$. Let us introduce the following dimensionless parameters

$$\bar{x} = \frac{x}{d}; \quad \bar{t} = t \cdot \frac{v_0}{d}; \quad \bar{v} = \frac{v}{v_0}; \quad \bar{S} = \frac{S}{S_0} = \frac{S}{(\pi d^2/4)};$$

$$\bar{R}_h = \frac{R_h}{d} = 0.25 \cdot (1 - \sin\phi/\phi) \quad \text{(see Eq. (3.46))}; \quad \bar{C}_{Sh} = \frac{C_{Sh}}{\sqrt{g}}.$$

With new variables the system of equations describing gravity fluid flow transforms to

$$\begin{cases} \dfrac{\rho_0 S_0 v_0}{d} \cdot \left(\dfrac{\partial \bar{S}}{\partial \bar{t}} + \dfrac{\partial \bar{v} \bar{S}}{\partial \bar{x}} \right) = 0 \\ \dfrac{\rho_0 {v_0}^2 S_0}{d} \cdot \left[\dfrac{\partial \bar{v} \bar{S}}{\partial \bar{t}} + \dfrac{\partial}{\partial \bar{x}} \left(\bar{v}^2 \bar{S} + \dfrac{gd \cos \alpha}{{v_0}^2} \displaystyle\int_0^{\bar{S}} \bar{S} \, d\bar{h} \right) \right] \\ \qquad = -\rho_0 g S_0 \bar{S} \sin \alpha - \dfrac{\rho_0 g \cos \alpha \cdot S_0 {v_0}^2}{gd} \cdot \dfrac{\bar{S} \cdot \bar{v}|\bar{v}|}{\bar{R}_h \bar{C}_{Sh}} \end{cases}$$

or, after simplification

$$\begin{cases} \dfrac{\partial \bar{S}}{\partial \bar{t}} + \dfrac{\partial \bar{v} \bar{S}}{\partial \bar{x}} = 0 \\ \dfrac{\partial \bar{v} \bar{S}}{\partial \bar{t}} + \dfrac{\partial}{\partial \bar{x}} \left(\bar{v}^2 \bar{S} + \dfrac{gd \cos \alpha}{{v_0}^2} \displaystyle\int_0^{\bar{S}} \bar{S} \, d\bar{h} \right) = -\dfrac{gd \sin \alpha}{{v_0}^2} \bar{S} \cdot - \dfrac{\bar{v}|\bar{v}| \cdot \bar{S} \cos \alpha}{\bar{R}_h \bar{C}_{Sh}} \end{cases} \quad (8.6)$$

It is seen that, in the system of equations (8.6), besides the dimensionless angle α ($\Pi_1 = \alpha$) there is one more dimensionless complex $\Pi_2 = gd/{v_0}^2$ or $v_0/\sqrt{gd}$. We already met the latter one in Section 7.4; it is the so-called Froude number

$$Fr = \frac{v_0}{\sqrt{gd}}. \quad (8.7)$$

Thus, the Froude similarity criterion (Froude number) is obtained in the mathematical problem on gravity fluid flow in a pipeline, which was earlier derived on the basis of general reasoning on the similarity of such flows.

8.4 Pipeline Transportation of Oil Products. Batching

At the present time light oil products (benzenes, kerosenes, diesel fuels and others) are pumped by the *batching method* (Ishmuchamedov *et al.*, 1999). In oil treatment the factory plants simultaneously produce a plethora of oil products, mostly the so-called light oil products and, above all, motor oils. As a result of further compounding (mixing of two or several oil products to prepare fuels with given properties) there are obtained different sorts of oil products ready for use. It is evident that the building of a separate pipeline for each of produced oil products would be unprofitable, therefore most of the oil products are transported by one and the same pipeline, pumping them in series one after another (batching).

8.4.1
Principle of Oil Product Batching by Direct Contact

The essence of batching by direct contact consists in the different oil products being combined in separate batches, each of several thousand or even ten thousand tons, which are pumped into the pipeline in series, one after another, and transported to the user. In this way each batch displaces the previous one and is, in turn, displaced by the following batch. It is as if the oil-pipeline along its full length were filled by different oil products arranged in a chain and contacting with each other at the places where one batch comes to an end and another begins. Thus, the key advantage in the batching of oil products is that different sorts of oil products are pumped *not along different pipes but along one and the same pipe.*

However, despite all the advantages of batching it has one significant disadvantage consisting in the formation of a mixture of different oil products by their mutual displacement in the pipeline. Although the mixing *of similar oil products,* for example benzenes of different sorts or diesel fuels of different sorts, threatens little the quality of the resulting oil product, because oil products relating to one group of fuels are more compatible than oil products relating to different groups, the mixing of dissimilar oil products, for example benzenes, kerosenes and diesel fuels threatens significantly the quality of the oil products. Nevertheless, batching with direct contact of the oil products has received wide recognition because the quantity of mixture forming in the contact zones of batches moving in series is relativey small compared to the large volume of transported fuels and the whole mixture can be decomposed into the initial oil products, preserving the quality of the latter.

The mixture formed in the contact zone of oil products is caused by physical processes inherent to fluid flow in the pipeline and to displacment of one oil product by another. If contacting oil products displaced each other as rigid bars with plane interface boundaries, mixing in the contacting zone would of course be absent. However, fluid oil products are not rigid bodies and mutual displacement happens *nonuniformly* over the pipeline cross-section. The velocities of fluid particles at different points of the pipeline cross-section are distinct. At the pipeline wall they vanish whereas at the pipeline axis they achieve a maximum value. Thus the displacement of one oil product by another occurs more at the pipeline center than at the pipeline wall. At each instance of time the wedge behind the moving oil product becomes as if penetrated into the leading fluid, the penetration happening more when the profile of the average velocity is more stretched along the pipeline axis. There takes place so-called *convection* (or *convective diffusion*) of the impurity of one oil product into another one owing to, and together with, fluid layers transferring relative to each other.

Nonuniformity of the fluid average velocity distribution at the pipeline cross-section is not the only reason for mixture formation of oil products in the zone of their contact. Light oil products are, as a rule, pumped in a turbulent flow regime in which the fluid particles do not move parallel the pipeline walls but execute chaotic turbulent motion, as can be seen in smoke jets gushing from heat plants. In turbulent flows there exists intensive mixing of different particles over the pipeline cross-section caused by velocity fluctuations and the chaotic motion of particles. This process is called *turbulent diffusion*. It mixes over the pipeline cross-section the edge of the displacing fluid as well as the rest of the fluid to be displaced, causing their more or less uniform distribution in each pipeline cross-section.

Hence, the mixing process of oil products displacing and to be displaced happens in accordance with the following scheme: the edge of the oil product moving behind penetrates the oil product moving in front while processes of turbulent diffusion mix the penetrated impurity over the pipeline cross-section. Since the concentration of the displacing oil product at the pipeline axis is greater than at the wall, transport of the displacing oil product into the region occupied by oil product to be displaced occurs. Conversely, back transport of the oil product to be displaced into the region of the displacing oil product also occurs. These processes are inseparable. They operate permanently and simultaneously over the course of the displacement time, determining thus the intensity of the *longitudinal mixing*, the volume and length of the resulting mixture.

8.4.2
Modeling of Mixture Formation in Oil Product Batching

Light oil products possess the following property: if a volume V_1 of the first oil product is mixed with a volume V_2 of the second oil product, the volume V_c of the resulting mixture is, to a high degree of accuracy, equal to the sum of the volumes of the components $V_c = V_1 + V_2$. Therefore, the additivity property of fluid volume on mixing of its components is taken as *a main assumption* in the construction of the model of mixture formation in a pipeline.

If we denote through ρ_1 and ρ_2 the densities of the contacting oil products, the volume concentrations θ_1 and θ_2 of the oil products may be expressed through these quantities and the density ρ_c of the mixture. In accordance with the mass conservation law we have

$$\rho_1 V_1 + \rho_2 V_2 = \rho_c V_c.$$

From this it follows that

$$\rho_1 \frac{V_1}{V_c} + \rho_2 \frac{V_2}{V_c} = \rho_c \Rightarrow \rho_1\theta_1 + \rho_2\theta_2 = \rho_c.$$

Taking further that $\theta_2 = 1 - \theta_1$ or $\theta_1 = 1 - \theta_2$ we obtain the following formulas

$$\theta_1 = \frac{\rho_c - \rho_2}{\rho_1 - \rho_2}, \quad \theta_2 = \frac{\rho_c - \rho_1}{\rho_2 - \rho_1}. \tag{8.8}$$

We now develop the model of mixture formation in the fluid flow in pipeline. The mixture of oil products in the contact zone can be characterized by the concentration $\theta(x, t)$ of one of the oil products, for example the displacing one, in one-dimensional fluid flow in a pipeline. The cases $\theta \neq 1$ or $\theta \neq 0$ have a direct relationship to the mixture. In the mixture region $0 < \theta < 1$. The case $\theta = 1$ corresponds to the region of displacing oil product, whereas the case $\theta = 0$ corresponds to the oil product ahead that is to be displaced.

In order to derive a mathematical model of mixture formation it is necessary to reveal the mass exchange regularities of oil products in the mixture region, that is to specify the relation between the volumetric flow rate $q(x, t)$ of the displacing oil product (the volumetric flow rate of the oil product to be displaced would clearly be equal to $v_0 S - q(x, t)$) and the parameters of the concentration distribution $\theta(x, t)$ in the flow.

Figure 8.1 represents a scheme of mass exchange in an arbitrary cross-section of the mixture region. The total fluid flow rate through the cross-section x in the moving frame of reference is equal to zero, but the transfers of fluid (mixture of oil products) from left to right and from right to left are non-zero; they are equal in magnitude but opposite in sign. The flow of the mixture through the cross-section x from left to right with flow rate w_1 happens mainly in the central part of the pipeline, while the flow of the mixture in the reverse direction from right to left with flow rate $w_2 = -w_1$ occurs chiefly close to the internal surface of the pipeline.

The flow rate $w = w_1 = -w_2$ is determined by the profile of the velocity $\hat{u}(r)$ averaged over the interval $0 \leq r \leq r_*$

$$w = 2\pi \cdot \int_0^{r_*} r \cdot [\hat{u}(r) - v_0]\, dr. \tag{8.9}$$

If we take the velocity profile $\hat{u}(r)$ equal to the logarithmic profile in turbulent flow (Loitzyanskiy, 1987)

$$\frac{\hat{u}(r) - u_{\max}}{u_*} = \frac{1}{\kappa} \cdot \ln\left(1 - \frac{r}{r_0}\right) \tag{8.10}$$

from Eqs. (8.9) and (8.10) ensue the following relations

$$v_0 = u_{\max} - 4.08 \cdot u_*, \quad r_* = 0.805 \cdot r_0, \quad w = 1.26 \cdot u_* \cdot S. \tag{8.11}$$

Here $\kappa \cong 0.4$ is the Karman constant and u_* is the *dynamic velocity*, see Section 3.5. The latter is expressed through the tangential frictional stress $|\tau_w|$ at the pipeline wall

$$|\tau_w| = \rho \cdot u_*^2, \quad u_* = \sqrt{\frac{|\tau_w|}{\rho}}.$$

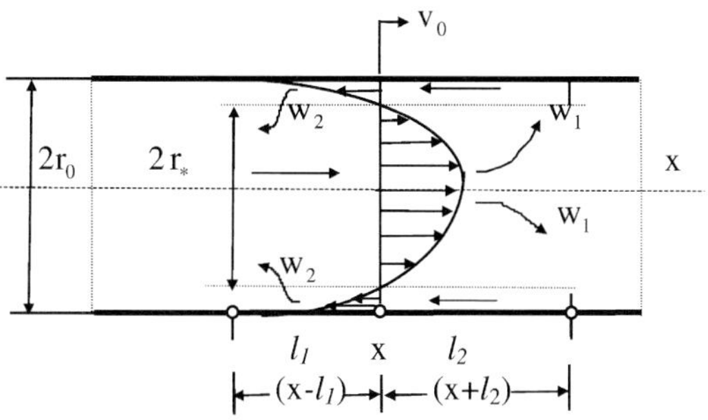

Figure 8.1 Scheme of mass exchange in the mixture region.

Since $|\tau_w| = \lambda/8 \cdot \rho v_0^2$, where λ is the hydraulic resistance factor, the dynamic velocity is related to this factor by

$$u_* = \sqrt{\frac{\lambda}{8}} \cdot v_0.$$

Insertion of the expression for u_* in Eq. (8.11) yields the connection between the mass exchange flow rate w and the pumping flow rate $Q = v_0 S$

$$w = 1.26 \cdot \sqrt{\frac{\lambda}{8}} \cdot v_0 S = 0.446 \cdot \sqrt{\lambda} \cdot Q. \tag{8.12}$$

From this formula it follows that the quantity exchange flow rates are relatively not large. So, for example, at $\lambda = 0.022$ the quantity $w = 0.066\, v_0 S$, which is only 6.6% of the pumping flow rate.

Counter flows of fluid transfer both the first and second oil product through the cross-section x of the moving frame of reference, but the average concentrations θ' and θ'' in the transfer flows are different. Thus, the flow rate $q(x, t)$ of the displacing oil product through the cross-section x is given by

$$q(x, t) = w \cdot \theta' - w \cdot \theta'' = w \cdot (\theta' - \theta'').$$

In the first flux (from left to right) the concentration θ' is equal to the concentration of the displacing oil product averaged over the cross-section at some distance l_1 behind the cross-section x. In the second flux (from right to left) the concentration θ'' is equal to the concentration of the displacing oil product at some distance l_2 ahead of the cross-section x. Lengths l_1 and l_2 can be called mixing lengths, since they are equal to the lengths over which the turbulent diffusion mixes the invading impurity over the pipeline cross-section.

Accurate to small quantities of the highest order it could be written as

$$\begin{cases} \theta' = \theta(x - l_1, t) \cong \theta(x, t) - l_1 \cdot \dfrac{\partial\theta}{\partial x} + \ldots, \\ \theta'' = \theta(x + l_2, t) \cong \theta(x, t) + l_2 \cdot \dfrac{\partial\theta}{\partial x} + \ldots \end{cases} \tag{8.13}$$

and

$$q(x,t) = w\cdot\theta' - w\cdot\theta'' = w\cdot(\theta' - \theta'') = -w\cdot(l_1 + l_2)\cdot\frac{\partial\theta}{\partial x}.$$

Substitution of w from (8.12) in the latter relation yields

$$q(x,t) = -0.446\cdot\sqrt{\lambda}\cdot(l_1 + l_2)\cdot\frac{\partial\theta}{\partial x}\cdot v_0 S$$

or

$$q(x,t) = -K\cdot\frac{\partial\theta}{\partial x}\cdot S \tag{8.14}$$

where

$$K = 0.446\cdot\sqrt{\lambda}\cdot(l_1 + l_2)\cdot v_0. \tag{8.15}$$

The relation (8.14) expressing the proportionality of the volumetric flow rate $q(x,t)$ of the displacing oil product gradient $\partial\theta/\partial x$ to its concentration is called *the law of longitudinal mixing* and the factor K ($m^2\ s^{-1}$) the *effective factor of the longitudinal mixing*. The minus sign in Eq. (8.14) shows that the flux of each oil product is directed from the higher concentration to the lower one, that is opposite to the concentration gradient of the oil product under consideration.

There are many theoretical and experimental formulas for the factor K of longitudinal mixing (Ishmuchamedov *et al.*, 1999). We consider one of them, namely the formula derived by Taylor when investigating the dispersion of an impurity in turbulent fluid flow in a pipe

$$K = 1.785\cdot\sqrt{\lambda}\cdot v_0 d. \tag{8.16}$$

Comparing this formula with Eq. (8.15), we find for the sum of the mixing lengths $(l_1 + l_2)$ the value $4d$. This result appears to be true for Reynolds numbers higher than $3\cdot 10^4$.

8.4.3 Equation of Longitudinal Mixing

To derive a mathematical model describing the process of mixture formation in a pipeline for displacement of one fluid by another one we use the volume balance equation for each component. Since the sum of the volume concentrations θ_1 and θ_2 of oil products is equal to 1, it is enough to write only one balance equation for one of the components, for example the displacing component, taking $\theta_1 = \theta(x,t)$, $\theta_2 = 1 - \theta(x,t)$.

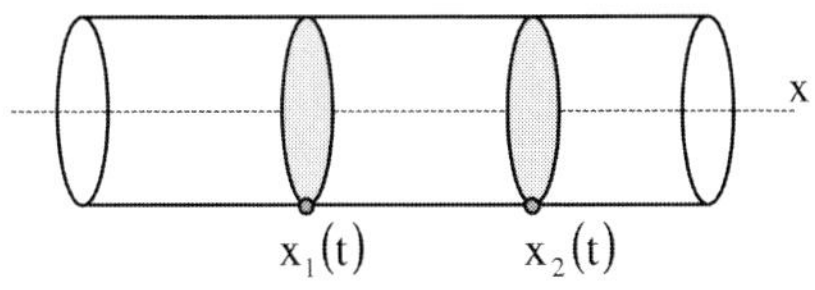

Figure 8.2 Derivation of the volume balance equation.

Consider a *fluid volume* (see Section 1.2) enclosed between cross-sections $x_1(t)$ and $x_2(t)$, Fig. 8.2.

The first (displacing) oil product occupies only a part of this volume V_1 which can be written as

$$V_1 = \int_{x_1(t)}^{x_2(t)} \theta(x, t) \cdot S\,\mathrm{d}x.$$

If the exchange of oil products through the cross-sections $x_1(t)$ and $x_2(t)$ were absent, the quantity V_1 would be constant, but in reality it is not. The quantity V_1 varies owing to mutual penetration of the oil products into each other and this variation is determined by the difference in transfer flow rates $q(x_1, t) - q(x_2, t)$. Thus

$$\frac{\mathrm{d}V_1}{\mathrm{d}t} = \frac{\mathrm{d}}{\mathrm{d}t}\left(\int_{x_1(t)}^{x_2(t)} \theta(x, t) \cdot S\,\mathrm{d}x\right) = q(x_1, t) - q(x_2, t). \tag{8.17}$$

If we take into account the following identities:

$$\frac{\mathrm{d}}{\mathrm{d}t}\left(\int_{x_1(t)}^{x_2(t)} \theta(x, t) \cdot S\,\mathrm{d}x\right) = \int_{x_1(t)}^{x_2(t)} \left(\frac{\partial \theta S}{\partial t} + \frac{\partial v_0 \theta S}{\partial x}\right) \mathrm{d}x, \text{ see Eq. (1.4)},$$

$$q(x_1, t) - q(x_2, t) = -\int_{x_1(t)}^{x_2(t)} \frac{\partial q}{\partial x}\,\mathrm{d}x,$$

the relation (8.17) may be written as

$$\int_{x_1(t)}^{x_2(t)} \left(\frac{\partial \theta S}{\partial t} + \frac{\partial v_0 \theta S}{\partial x}\right) \mathrm{d}x = -\int_{x_1(t)}^{x_2(t)} \frac{\partial q}{\partial x}\,\mathrm{d}x.$$

If, in addition, we recall that the *fluid volume* under consideration is arbitrarily chosen, that is the integration limits $x_1(t)$ and $x_2(t)$ are arbitrarily chosen, then from the latter integral equation follows the differential equation

$$\frac{\partial \theta S}{\partial t} + \frac{\partial v_0 \theta S}{\partial x} = -\frac{\partial q}{\partial x}$$

expressing the volume balance of the first oil product in the mixture.

Insertion instead of q, using its expression through the concentration gradient $q(x, t) = -K \cdot S \cdot \partial\theta/\partial x$ (see Eq. (8.14)) with regard to the conditions $S = \text{const.}$ and $v_0 = \text{const.}$ yields the differential equation for the concentration

$\theta(x, t)$ of the displacing oil product in the mixture

$$\frac{\partial\theta}{\partial t} + v_0\frac{\partial\theta}{\partial x} = -K \cdot \frac{\partial^2\theta}{\partial x^2} \tag{8.18}$$

Equation (8.18) represents the *differential equation of longitudinal mixing* referring to the class of *heat conduction equations* (Ishmuchamedov *et al.*, 1999).

If we introduce the dimensionless variables $\bar{x} = x/L$ and $\bar{t} = t/(L/v_0)$, Eq. (8.18) can be rewritten in dimensionless form

$$\frac{1}{(L/v_0)} \cdot \frac{\partial\theta}{\partial\bar{t}} + \frac{v_0}{L} \cdot \frac{\partial\theta}{\partial\bar{x}} = \frac{K}{L^2} \cdot \frac{\partial^2\theta}{\partial\bar{x}^2}$$

or

$$\frac{\partial\theta}{\partial\bar{t}} + \frac{\partial\theta}{\partial\bar{x}} = Pe^{-1} \cdot \frac{\partial^2\theta}{\partial\bar{x}^2} \tag{8.19}$$

where the dimensionless parameter $Pe = v_0L/K$ called the *Peclet number* is the main characteristic of quantity of mixture forming in the course of oil product pumping and the similarity criterion in problems on longitudinal mixing of fluids by their displacement in the pipe.

8.4.4
Self-Similar Solutions

The reasoning of dimensional theory can sometimes bring very important results, being capable of finding solutions of differential equations reflecting the most important peculiarities of processes and being in some sense *limiting* for solutions, taking into account the minor peculiarities and details of processes. To such solutions belong the so-called *self-similar* solutions. Let us demonstrate these with the example of Eq. (8.18).

In problems on the displacement of one oil product by another we can consider the dimensional parameters x, t, K, v_0, L. However, the number of these parameters could be reduced without serious consequences for the physics of the processes. If for example we change to a frame of reference $x_* = x - v_0t$ moving with average velocity v_0 of the pumping fluid, then Eq. (8.18) would be simplified to

$$\frac{\partial\theta}{\partial t} = K \cdot \frac{\partial^2\theta}{\partial {x_*}^2} \tag{8.20}$$

If the boundary conditions at the pipeline section edges have become, after frame of reference transformation, movable $x_* = -v_0t$ and $x_* = L - v_0t$, but far from the region occupied by the mixture they can be replaced by conditions at "infinitiy" ($\pm\infty$), that is $\theta \to 0$ when $x \to +\infty$ and $\theta \to 1$ when $x \to -\infty$

and the initial conditions at $t = 0$ we can take as *instantaneous* changing of oil products, namely $\theta(x_*, 0) = 1$ at $x_* < 0$ and $\theta(x_*, 0) = 0$ at $x_* > 0$ or

$$\theta(x_*, 0) = \begin{cases} 1, & x_* < 0, \\ 0, & x_* > 0. \end{cases} \tag{8.21}$$

Then, in the problem under consideration, there remain only three dimensional quantities x, t, K. Since the solution of the problem is the dimensional function θ, depending on three dimensionally-independent parameters from which may be built *only one* dimensionless combination then

$$\xi = \frac{x_*}{\sqrt{Kt}} \tag{8.22}$$

in accordance with the Π-theorem the solution should depend on one variable. Thus, the solution of the problem (8.20) and (8.21) has to be sought in the form $\theta(x_*, t) = \theta(\xi)$.

Such a solution is called *self-similar* because it is as if it transforms itself in space in a similar way, namely the solution of the problem at an arbitrary instant of time t may be obtained from the solution of the problem at any previous instant of time t_1 by stretching the distribution $\theta(x_*, t_1)$ along the x_*-axis with factor $\sqrt{K \cdot t/t_1}$, because it is well known that the multiplication of the function argument by a certain number brings extension or compression of the function graph along the abscissa axis.

Substitution of $\theta(x_*, t) = \theta(x_*/\sqrt{Kt})$ in Eq. (8.20) leads to an ordinary differential equation of the second order

$$-\frac{\xi}{2} \cdot \frac{d\theta}{d\xi} = \frac{d^2\theta}{d\xi^2}$$

from which follows

$$\theta(\xi) = A \cdot \int_0^{\xi} e^{-\alpha^2/4}\, d\alpha + B$$

where A and B are constants of integration.

Using the boundary conditions $\theta \to 0$ at $\xi \to +\infty$ and $\theta \to 1$ at $\xi \to -\infty$ gives $A = 1/(2\sqrt{\pi})$, $B = 1/2$. In deriving the latter we use the known equality

$$\int_0^{\infty} e^{-\alpha^2}\, d\alpha = \frac{\sqrt{\pi}}{2}.$$

The solution of the problem is

$$\theta(\xi) = \frac{1}{2}\left(1 - \frac{2}{\sqrt{\pi}} \cdot \int_0^{\xi/2} e^{-\alpha^2}\, d\alpha\right)$$

or in dimensional variables

$$\theta(x_*, t) = \frac{1}{2}\left(1 - \frac{2}{\sqrt{\pi}}\int_0^{x_*/\sqrt{4Kt}} e^{-\alpha^2}\, d\alpha\right). \tag{8.23}$$

Graphs of the concentration distribution $\theta(x_\bullet, t)$ in the moving frame of reference are shown in Fig. 8.3. The heavy line depicts the initial distribution whereas other lines show the concentration distribution at successively increased instants of time.

A remark on the imperfection of the model. It can be seen at once that the obtained solution has a defect. This is that the mixture, which initially was absent, at just the next instant of time would propagate through the whole pipeline length. In fact such a case could not occur in practice and the obtained paradox is the result of imperfectness of the model. The dispersion model of longitudinal mixing, as in general all models of this kind, represents only a certain schematization, in the given case it is the formation of a mixture in the contact zone of the displacing fluid and the fluid to be displaced. But the result of such a schematization does not give a particularly bad result. The function $\theta(x, t)$ tends very quickly to zero at $x \to \infty$ and to one at $x \to -\infty$. In the main domain of variability this function approximates the concentration distribution in the mixture zone well. Therefore, the dispersion model of longitudinal mixing based on Eq. (8.20) has received wide application in calculations of mixtures forming in the batching of oil products.

Mixture volume. The volume V_c of a mixture of pumping oil products forming in a pipeline at the instant of time t calculated by Eq. (8.23) in the range of concentrations $0.01 < \theta < 0.99$ is determined by the expression

$$V_c \cong 6.58 \cdot \frac{\pi d^2}{4} \cdot \sqrt{K \cdot t}. \tag{8.24}$$

A remark on the quantity of mixture volume. Taylor obtained Eq. (8.16) for the factor K of longitudinal mixing under the assumption that the densities and viscosities of contacting fluids are close to each other. Therefore, strictly speaking, this equation is not suitable for the case when the densities and viscosities of contacting fluids differ significantly from each other. Such situations occur in the pumping of oil products, for example

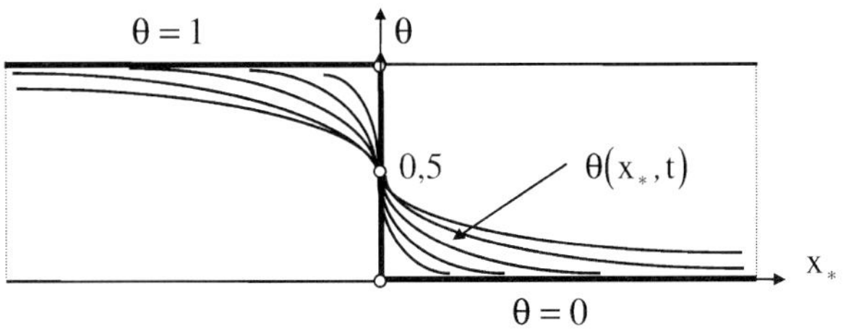

Figure 8.3 Self-similar distribution of concentration.

when pumping benzenes ($\rho \approx 730$–750 kg m^{-3}, $\nu \approx 0.6$ cSt) with diesel fuels ($\rho \approx 830$–850 kg m^{-3}, $\nu \approx 4$–9 cSt). It is evident that for these fluids the factors of hydraulic resistance λ are also different.

The formula for the mixture volume V_c as applied to the case under consideration can be improved if we calculate the mixture volume as the arithmetical mean of two volumes: the first calculated by Eq. (8.24) with the factor $K_1 = 1.785 \cdot \sqrt{\lambda_1} \cdot v_0 d$ and the second by Eq. (8.24) with the factor $K_2 = 1.785 \cdot \sqrt{\lambda_2} \cdot v_0 d$. It is equivalent to formula (8.24) if we take

$$K = 0.446 \cdot \left(\sqrt[4]{\lambda_1} + \sqrt[4]{\lambda_2}\right)^2 \cdot v_0 d. \tag{8.25}$$

Exercise. It is required to calculate the length and the volume of the mixture region in a symmetric concentration range $0.01 < \theta < 0.99$ when batching benzene ($\nu_B = 0.6$ cSt) and diesel fuel ($\nu_D = 6$ cSt) in an oil-pipeline ($D = 530 \times 8$ mm, $\Delta = 0.15$ mm, $L = 700$ km) when the transportation of oil products occurs with flow rate $Q = 1000$ m^3 h^{-1}.

Solution. Determine the average transportation velocity v_0 of the oil products

$$v_0 = \frac{4Q}{\pi d^2} = \frac{4 \cdot 1000/3600}{3.14 \cdot (0.530 - 2 \cdot 0.008)^2} \cong 1.34 \text{ m s}^{-1}.$$

Calculate the Reynolds numbers

$$Re_B = \frac{v_0 d}{\nu_B} = \frac{1.34 \cdot 0.514}{0.6 \cdot 10^{-6}} \cong 1\,147\,933,$$

$$Re_D = \frac{v_0 d}{\nu_D} = \frac{1.34 \cdot 0.514}{6 \cdot 10^{-6}} \cong 114\,793.$$

Calculate the hydraulic resistance factors λ_B and λ_D

$$\lambda_B = 0.11 \cdot \left(\frac{0.15}{514} + \frac{68}{1\,147\,933}\right)^{0.25} \cong 0.015,$$

$$\lambda_D = 0.11 \cdot \left(\frac{0.15}{514} + \frac{68}{114\,793}\right)^{0.25} \cong 0.019.$$

Calculate with Eq. (8.25) the factor K of longitudinal mixing

$$K = 0.446 \cdot \left(\sqrt[4]{0.015} + \sqrt[4]{0.019}\right)^2 \cdot 1.34 \cdot 0.514 \cong 0.222 \text{ m}^2 \text{ s}^{-1}.$$

With Eq. (8.24) calculate the mixture volume V_c

$$V_c \cong 6.58 \cdot \frac{\pi d^2}{4} \cdot \sqrt{K \cdot t} = 6.58 \cdot \frac{\pi d^2}{4} \cdot \sqrt{K \cdot \frac{L}{v_0}}$$

$$= 6.58 \cdot \frac{3.14 \cdot 0.514^2}{4} \cdot \sqrt{0.222 \cdot \frac{700\,000}{1.34}} \cong 465 \text{ m}^3.$$

The length of the mixture region l_c is

$$l_c = \frac{V_c}{\pi d^2/4} = \frac{465}{3.14 \cdot 0.514^2/4} \cong 2242 \text{ m or } 2.242 \text{ km}.$$

Answer: 2242 m; 465 m^3.

It should be noted that non-self-similar solutions of the considered problem, that is solutions taking into account the finite extent of the pipeline and the conditions at its edges, differ slightly from the self-similar solution. The facility to formulate the problem, making it self-similar and yet without changing its most important features, characterizes the high skill of the scientist. That is why self-similar solutions play such an important role in the different fields of science (Sedov, 1965).

References

Archangelskiy V.A., (**1947**) *Calculation of Non-Stationary Flow in Open Water Courses*, Academy of Sciences USSR, Moscow (in Russian).

Charniy I.A., (**1975**) *Non-Stationary Motion of Real Fluid in Pipes*, 2nd edn., Nedra, Moscow (in Russian).

Christianowitch S.A., (**1938**) Non-Stationary Flow in Channels and Rivers, *Collected articles on Some Problems in Continuum Mechanics*, Academy of Sciences USSR, Moscow (in Russian).

Dodge D.W., Metzner A.B., (**1958**) Turbulent Flow of Non-Newtonian Systems, *AIChE J.*, **2**, 189–204 .

Ginsburg I.P., (**1958**) *Applied Hydro-Gas-Dynamics*, LGU, Leningrad (in Russian).

Ishmuchamedov I.T., Isaev S.L., Lurie M.V., Makarov C.P., (**1999**) *Pipeline Transportation of Oil Products*, Oil and Gas, Moscow (in Russian).

Leibenson L.S., Vilker D.S., Shumilov P.P., Yablonskiy V.S., (**1934**) *Hydraulics*, 2nd edn., Gosgorgeolnephteizdat, Moscow, Leningrad (in Russian).

Loitzyanskiy L.G., (**1987**) *Mechanics of Fluid and Gas*, Nauka, Moscow (in Russian).

Lurie M.V., (**2001**) Technique of Scientific Researches. Dimensionality, Similarity and Simulation of Phenomena in Problems of Oil Transportation and Oil- and Gas-Storage, Oil and Gas, Moscow (in Russian).

Lurie M.V., Polyanskaya L.V., (**2000**) About One Dangerous Source of Hydraulic Shock Waves in Oil and Oil Products, *Oil Facilities*, No. 8 (in Russian).

Lurie M.V., Podoba N.A., (**1984**) Modification of Karman Theory for Turbulent Shear Flows, *Papers of the Academy of Science of the USSR*, **279**(3), (in Russian).

Vasil'ev G.G., Korobkov G.E., Lurie M.V. *et al.*, (**2002**) *Pipeline Transportation of Oil*, Vol.1, S.M. Veinstock, Nedra (in Russian).

Polyanscaya L.V., (**1965**) Investigation of non-stationary processes in changing operation regime with centrifugal pumps, Kand. Sci. Thesis, Gubkin Oil & Gas Institute, Moscow (in Russian).

Porshakov Yu.P., Kosachenko A.N., Nikishin V.I., (**2001**) *Power of Pipeline Gas Transportation*, Oil and Gas, Moscow (in Russian).

Potapov A.G., (**1975**) *Hydraulic Resistance Factor in Turbulent Flow of Viscous-Plastic Fluids*, Volgogradnipinepht, Volgograd, No. 23 (in Russian).

Rozhdestvenskiy B.L., Yanenko N.N., (**1977**) *Systems of Quasi-Linear Differential Equations*, Nauka, Moscow (in Russian).

Romanova N.A., (**1985**) *Laminar and Turbulent Flows in Pipes and Channels with Moving Walls*, Kand. Sci. Thesis, Gubkin Oil & Gas Institute, Moscow (in Russian).

Samarskiy A.A., (**1977**) *Introduction to the Theory of Difference Schemes*, Nauka, Moscow (in Russian).

Sedov L.I., (**1965**) *Methods of Similarity and Dimensionality in Mechanics*, Nauka, Moscow (in Russian).

Tihonov A.N., Samarskiy A.A., (**1966**) *Equations of Mathematical Physics*, Nauka, Moscow (in Russian).

Wilkenson U.L., (**1960**) *Non-Newtonian Fluids*, Pergamon Press, London, NewYork.

Modeling of Oil Product and Gas Pipeline Transportation. Michael V. Lurie

ISBN: 978-3-527-40833-7

Appendices

Appendix A. Increment and Differential of a Function

In some sections of this book the formula for the increment Δf of the differentiable $f(x)$

$$f(x+\Delta x)-f(x)\approx \left.\frac{df}{dx}\right|_x\cdot \Delta x$$

where Δx is the increment of the argument. Let us explain this formula.

Figure A.1 shows a graph of a differentiable and, consequently, continuous function $y=f(x)$. The difference $f(x+\Delta x)-f(x)$ of the values of this function at two points x and $(x+\Delta x)$ is called the increment of the function Δf.

If at the point M on the abscissa x we draw a tangent to the plot of the function, the increment of the function can be interpreted as the sum of the two segments AB and BC, Fig. A2.

Since

$$\frac{df}{dx}=\lim_{\Delta x\to 0}\frac{f(x+\Delta x)-f(x)}{\Delta x}$$

this means that

$$\left.\frac{df}{dx}\right|_x-\frac{f(x+\Delta x)-f(x)}{\Delta x}=\varepsilon\to 0\quad \text{at}\quad \Delta x\to 0$$

that is, ε is an infinitesimal quantity. From this it follows that the increment $\Delta f=(x+\Delta x)-f(x)$ of a differentiable function $f(x)$ can be represented as

$$f(x+\Delta x)-f(x)=\underbrace{\left.\frac{df}{dx}\right|_x\cdot \Delta x}_{\text{AB}}+\underbrace{\varepsilon\cdot\Delta x}_{\text{BC}}.$$

Since the value of the derivative at the point x is equal to the slope of the function plot at this point to the horizontal, it is evident that the first term on the right-hand side of the last formula is nothing but the length of the segment AB. If $df/dx\neq 0$, it appears to be a quantity of the same order as Δx.

Modeling of Oil Product and Gas Pipeline Transportation. Michael V. Lurie

ISBN: 978-3-527-40833-7

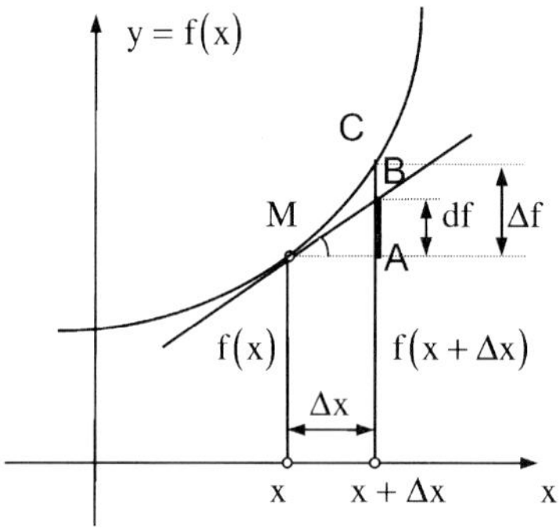

Figure A.1 Increment Δf and differential df of a function.

The second term $(\varepsilon \cdot \Delta x)$ is depicted in Fig. A.1 by segment BC. This term is a smaller quantity than the first term, because $\varepsilon \to 0$ at $\Delta x \to 0$. This means that the smaller Δx then the greater the accuracy when neglecting the second term. Thus, there is an approximation formula

$$f(x + \Delta x) - f(x) \approx \left.\frac{df}{dx}\right|_x \cdot \Delta x.$$

It is commonly said that this formula is true *up to an infinitesimal quantity of higher order.*

The principal linear (in $\Delta x \equiv dx$) part $\left.\frac{df}{dx}\right|_x \cdot \Delta x$ of the function increment Δf is called the function *differential* df

$$df = \left.\frac{df}{dx}\right|_x \cdot dx; \quad \Delta f = df + \varepsilon \cdot dx.$$

The differential of function df is plotted by segment AA, Fig. A.1.

Appendix B. Proof of the Formula

$$\frac{d}{dt}\int_{x_1(t)}^{x_2(t)} A(x,t) \cdot S(x,t)\,dx = \int_{x_1(t)}^{x_2(t)} \frac{\partial}{\partial t}[A(x,t) \cdot S(x,t)]\,dx$$
$$+ A(x,t) \cdot S(x,t)|_{x_2(t)} \cdot \frac{dx_2}{dt} - A(x,t) \cdot S(x,t)|_{x_1(t)} \cdot \frac{dx_1}{dt}$$

to calculate the *total derivative with respect to time of an integral with integration limits determining the motion of a fluid (individual) volume of a continuum in the pipeline.*

Proof. Consider two successive locations at the time instances t and $t + \Delta t$ of one and the same fluid volume in a pipeline, Fig. B.1.

In accordance with general definition of a derivative as a limit of the ratio of a function increment to an argument increment when the latter tends to zero

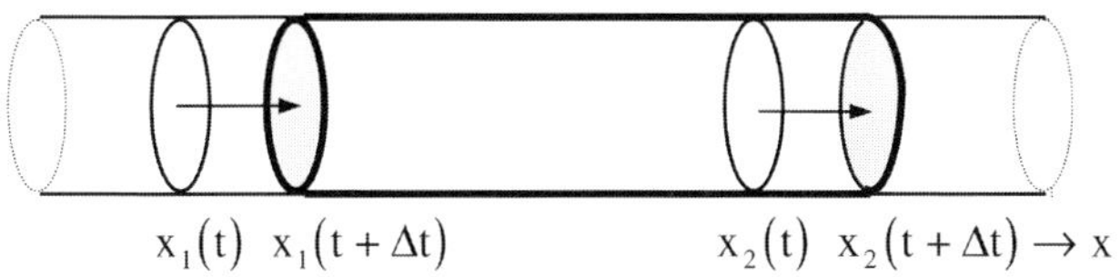

Figure B.1 Derivation of the formula of time differentiation of an integral with time dependent limits.

we have

$$\frac{\mathrm{d}}{\mathrm{d}t}\int_{x_1(t)}^{x_2(t)} A(x,t)\cdot S(x,t)\,\mathrm{d}x$$
$$= \lim_{\Delta t\to 0}\frac{1}{\Delta t}\left[\int_{x_1(t+\Delta t)}^{x_2(t+\Delta t)} A(x,t+\Delta t)\cdot S(x,t+\Delta t)\,\mathrm{d}x - \int_{x_1(t)}^{x_2(t)} A(x,t)\cdot S(x,t)\,\mathrm{d}x\right].$$

The integrals in square brackets can be represented as follows (see Fig. B.1)

$$\int_{x_1(t+\Delta t)}^{x_2(t+\Delta t)} A(x,t+\Delta t)\cdot S(x,t+\Delta t)\,\mathrm{d}x$$
$$= \int_{x_1(t+\Delta t)}^{x_2(t)} A(x,t+\Delta t)\cdot S(x,t+\Delta t)\,\mathrm{d}x + \int_{x_2(t)}^{x_2(t+\Delta t)} A(x,t+\Delta t)\cdot S(x,t+\Delta t)\,\mathrm{d}x$$

$$\int_{x_1(t)}^{x_2(t)} A(x,t)\cdot S(x,t)\,\mathrm{d}x$$
$$= \int_{x_1(t)}^{x_1(t+\Delta t)} A(x,t)\cdot S(x,t)\,\mathrm{d}x + \int_{x_1(t+\Delta t)}^{x_2(t+\Delta t)} A(x,t)\cdot S(x,t)\,\mathrm{d}x.$$

Consequently

$$\int_{x_1(t+\Delta t)}^{x_2(t+\Delta t)} A(x,t+\Delta t)\cdot S(x,t+\Delta t)\,\mathrm{d}x - \int_{x_1(t)}^{x_2(t)} A(x,t)\cdot S(x,t)\,\mathrm{d}x$$
$$= \int_{x_1(t+\Delta t)}^{x_2(t)} [A(x,t+\Delta t)\cdot S(x,t+\Delta t) - A(x,t)\cdot S(x,t)]\,\mathrm{d}x$$
$$+ \int_{x_2(t)}^{x_2(t+\Delta t)} A(x,t+\Delta t)\cdot S(x,t+\Delta t)\,\mathrm{d}x$$
$$- \int_{x_1(t)}^{x_1(t+\Delta t)} A(x,t)\cdot S(x,t)\,\mathrm{d}x.$$

Up to infinitesimal quantities of higher order this can be written as

$$\int_{x_1(t+\Delta t)}^{x_2(t)} [A(x, t+\Delta t)\cdot S(x, t+\Delta t) - A(x, t)\cdot S(x, t)]\,\mathrm{d}x$$

$$\cong \int_{x_1(t+\Delta t)}^{x_2(t)} \frac{\partial (A\cdot S)}{\partial t}\cdot \Delta t\cdot \mathrm{d}x$$

$$\int_{x_2(t)}^{x_2(t+\Delta t)} A(x, t+\Delta t)\cdot S(x, t+\Delta t)\,\mathrm{d}x \cong A(x_2, t)\cdot S(x_2, t)\cdot \frac{\mathrm{d}x_2}{\mathrm{d}t}\cdot \Delta t,$$

$$\int_{x_1(t)}^{x_1(t+\Delta t)} A(x, t)\cdot S(x, t)\,\mathrm{d}x \cong A(x_1, t)\cdot S(x_1, t)\cdot \frac{\mathrm{d}x_1}{\mathrm{d}t}\cdot \Delta t.$$

Dividing the sum of the latter three expressions by Δt and passing to the limit at $\Delta t \to 0$ we get the formula

$$\frac{\mathrm{d}}{\mathrm{d}t}\int_{x_1(t)}^{x_2(t)} A(x, t)\cdot S(x, t)\,\mathrm{d}x = \int_{x_1(t)}^{x_2(t)} \frac{\partial}{\partial t}[A(x, t)\cdot S(x, t)]\,\mathrm{d}x$$

$$+ A(x, t)\cdot S(x, t)|_{x_2(t)}\cdot \frac{\mathrm{d}x_2}{\mathrm{d}t} - A(x, t)\cdot S(x, t)|_{x_1(t)}\cdot \frac{\mathrm{d}x_1}{\mathrm{d}t}.$$

When going to the limit we took into account the continuity property of functions $x_1(t)$ and $x_2(t)$, namely $\lim_{\Delta t\to 0} x_1(t+\Delta t) = x_1(t)$ and $\lim_{\Delta t\to 0} x_2(t+\Delta t) = x_2(t)$.

Author Index

Modeling of Oil Product and Gas Pipeline Transportation. Michael V. Lurie

ISBN: 978-3-527-40833-7

Subject Index

Modeling of Oil Product and Gas Pipeline Transportation. Michael V. Lurie

ISBN: 978-3-527-40833-7